Times Table Tactics

Investigating multiplication facts

Peter Critchley

BEAM Education

BEAM Education is a specialist mathematics education publisher, dedicated to promoting the teaching and learning of mathematics as interesting, challenging and enjoyable.

BEAM materials cover teaching and learning needs from the age of 3 to 14. They deal with many of the classroom concerns voiced by teachers of mathematics, and offer practical support and help. BEAM materials include more than 80 publications, as well as a comprehensive range of mathematical games and equipment.

BEAM services include consultancy for companies, institutions and government, and a programme of courses and in-service training for schools, early years settings and local education authorities.

BEAM is an acknowledged expert in the field of mathematics education.

BEAM Education
Maze Workshops
72a Southgate Road
London N1 3JT

tel 020 7684 3323
fax 020 7684 3334
email info@beam.co.uk
www.beam.co.uk

Acknowledgements

The majority of the material in *Times Table Tactics* comes from the book *I learnt my tables but I don't know them* written by Peter Critchley for Suffolk LEA.

Our thanks to the teachers and schools who trialled *Times Table Tactics*:
Adam Barber and Parson Street Primary School, Bedminster, Bristol
Lea Bowers, Adèle Markey and Bury and Whitefield Jewish Primary School, Bury, Lancashire
Mary Chapman, Patricia Quinn and Virgo Fidelis Prep School, London
CM Draper and Moor Park School, Ludlow, Shropshire
Helen Elis-Jones and the University of Wales, Bangor, Gwynedd
Diane Jackson and The Dale Primary School, Stockport, Greater Manchester
Sandra Lemming and Bradfields Secondary School, Chatham, Kent
Caroline Logan and Whitgreave Primary School, Wolverhampton, West Midlands
Christine Lumb and Fearnville Primary School, Bradford, West Yorkshire
Hilary Osborn and Cranford House School, Moulsford, Oxfordshire
Janet Parfitt and Balfour Junior School, Brighton, Sussex
Damian Railston and Langworthy Road CP School, Salford, Greater Manchester
Val Shaw and Rangefield Primary School, Bromley, Kent
Alison Thomas and Reed First School, Royston, Hertfordshire
Jane Towers and Elangeni Junior School, Amersham, Buckinghamshire

Published by BEAM Education

© BEAM Education 2003

All rights reserved

ISBN 1 903142 03 2
British Library Cataloguing-in-Publication Data
Data available

Edited by Alice Smith and Henrietta Preston

Designed by Roger Marks

Cover by Mayur Mistry

Printed in England by Latimer Trend & Company Ltd

Preface

Some years ago, having discovered I was a maths teacher, a nurse told me, "I never did like maths at school. I can remember my Dad making me learn my tables but I still don't know them". When I asked her what she meant, she replied, "Well, if you ask me what seven eights are, I still have to go 'one eight is eight, two eights are sixteen, three eights are twenty-four' and so on!"

The nurse is typical of many people who learned the times tables by rote and who now find it difficult to recall the product of two single-digit numbers. What is more worrying, however, is that if they can't remember the products, they don't have any reliable strategies to calculate them.

The idea of having to know all of the times tables can be daunting for children, but the reality need not be. It is important for children to appreciate two things: firstly, that they do not have to learn all the multiplication facts to know their times tables, and secondly, that knowing the tables by rote does not mean that you can apply them readily. Once you understand the relationship between numbers, and have a feel for how they work together, the multiplication facts pop into your mind, rather than you having to chant through a rhyme to retrieve the fact you want.

Children will find that it is very empowering to have multiplication bonds at their fingertips. And teachers will find that learning is more likely to be successful, secure and long-lasting if children's experience of multiplication tables extends beyond learning them by rote. *Times Table Tactics* addresses this issue by offering teachers numerous ideas that will give their classes a wider perspective and richer experience of multiplication.

The ultimate aim of this book is to enable children to respond quickly to multiplications up to 10×10. This is a little more challenging than expecting children to know their multiplication tables. This book should be a useful aid to reaching this goal.

Peter Critchley

Contents

Introduction	7
Short activities matrix	8
Long activities matrix	9

Short activities

Step on – learning about the multiplication grid	12
Multiple choice – practising multiples with the grid	12
Hands and fingers – learning the 5 times table using hands and fingers to help	13
Purple patches – exploring the relationship between a number and its multiples	13
Kangaroo jumps – thinking of multiplication as a series of jumps on a number line	14
Wacky windows – a look at the common multiples of the 4 and 6 times tables	14
Charting the waves – arranging numbers on a multiples chart	15
Money matters – a set of questions exploring commutativity	15
And after the break… – working out the next multiple in a times table	16
A pressing matter – a look at division with the OHP calculator	16
It's a mystery – a number quiz with the OHP calculator	17
Calculating questions – finding the missing digits in multiplication equations	17
Box clever – completing a multiplication puzzle	18
Chosen numbers – recognising numbers that appear in the multiplication grid	18
Up and down – a number puzzle to practise multiplication	19
Every which way – working with multiplication equations	19
The spying game – using multiplication equations to solve problems	20
Scrambled numbers – recognising multiplication equations from jumbled numbers	20
Hundreds and thousands? – working out how many products there are in the multiplication grid	21
Broken keys – using mental strategies and calculators creatively	21

Long activities

Checks and stripes – exploring number sequences on a 1–100 grid	24
Figure out the fours – looking at the 4 times table in depth	26
Seven days – exploring the relationship between multiplication and addition	28
Double this, halve that – exploring doubling and halving numbers in multiplication equations	30
Building blocks – analysing multiplications	32
Heading into the great unknown – using known multiplication facts to work out unknown ones	34
Spot the difference – exploring the commutativity of the 6 times table	36
Methodical thinking – comparing four different multiplication strategies	38
49 is the magic number – how to use number bonds to work out all the multiplication facts	40
Testing times – a selection of tests for individuals and the class	42
It all adds up – using partitioning to work out large multiplication equations	44
Alphabet soup – finding number patterns in the times tables	46
Factoring in pairs – finding factors and factor pairs	48
Equal shares – exploring the relationship between multiplication and division	50
The missing link – using the link between multiplication and division	52
All or nothing – an introduction to the division grid	54
Number chains – a number challenge using all known multiplication facts	56
Bite size pieces – practising multiplication by partitioning	58
Telephone numbers – games and activities to gain familiarity with times tables	60
ISBN numbers – a real-life application of multiplication	62

Resource sheets	65

Introduction

Times Table Tactics provides multiplication and division problems that require mental strategies to solve them. The teaching involves unpicking and making explicit these strategies, so that children are provoked into using them, and into developing new ones. Doubling and halving, for example, are invaluable tactics, and knowing that four threes is the same as three fours is precious information that drastically reduces what you need to learn.

How to use the book

Times Table Tactics presents twenty short and nineteen long activities, as well as a selection of tests. The short activities are designed to be part of a lesson and can take up to 30 minutes. The long activities are full lesson plans, without the oral/mental starter.

You don't need to start at the beginning of the book – use the contents list or matrices on pages 8 and 9 to choose an appropriate activity. Many of the activities are flexible – you can adapt them for different multiplication tables. For those children who already know their multiplication facts, you can use the activities for consolidation (to make sure), problem solving, or extending the application of their knowledge to larger numbers. For example, if children know the multiplication facts for the 10 and 5 times tables, and understand the distributive law, they can grapple with the 15 times table.

Vocabulary and discussion

Use a variety of vocabulary in mathematics lessons: as well as 'multiplication' and 'times', talk about 'lots of', 'groups of', 'product', 'factor', 'quotient' and 'multiple'. Much of the learning takes place in the discussion that happens before, during and after the activity, both between children and between child and teacher. During the lesson, encourage children to work collaboratively, discussing and sharing their own methods for tackling, solving and extending the activities they have been given. Suggestions for what to say to the children are included in the activities.

Record of progress and use of equipment

Keep a record of the products each child can instantly recall, and update it regularly. You can do this simply by marking up a multiplication grid for each of them. Children can also maintain these records for themselves. Have available resources such as number grids, number lines and multiplication grids, both for individuals to use and displayed in the classroom for all to see and refer to when necessary. Many of the activities would also work well with an interactive whiteboard. The lessons give suggestions for ways of recording on the board or OHP. Comments in the margin offer additional information.

Short activities

Objectives	Step on p12	Multiple choice p12	Hands and fingers p13	Purple patches p13	Kangaroo jumps p14	Wacky windows p14	Charting the waves p15	Money matters p15	And after the break… p16	A pressing matter p16	It's a mystery p17	Calculating questions p17	Box clever p18	Chosen numbers p18	Up and down p19	Every which way p19	The spying game p20	Scrambled numbers p20	Hundreds and thousands? p21	Broken keys p21
Recognise multiples of 2, 5 and 10	✕	✕	✕				✕	✕						✕	✕		✕			
Recognise multiples of 3 and 4	✕	✕	✕											✕	✕		✕			
Recognise multiples of 6, 7, 8 and 9			✕														✕			
Find simple common multiples				✕													✕			
Know multiplication facts for the 2, 5 and 10 times tables											✕		✕			✕	✕			
Know multiplication facts for the 3 and 4 times tables											✕		✕				✕			
Know multiplication facts for the 6, 7, 8 and 9 times tables											✕						✕			
Use doubling or halving			✕																	
Use known number facts to carry out simple multiplications and divisions											✕	✕								
Use the relationship between multiplication and division							✕													
Use closely related facts							✕													
Use factors, eg 8 × 12 = 8 × 4 × 3													✕							
Partition, eg (47 × 6 = (40 × 6) + (7 × 6))																				
Understand multiplication as repeated addition																				
Understand and use the principle of the commutative law																✕				
Develop calculator skills and use a calculator effectively										✕									✕	✕
Solve mathematical problems and puzzles											✕							✕	✕	
Recognise and explain patterns and relationships	✕		✕		✕	✕	✕	✕		✕									✕	
Recognise products						✕	✕													

Long activities

Checks and stripes	p24
Figure out the fours	p26
Seven days	p28
Double this, halve that	p30
Building blocks	p32
Heading into the great unknown	p34
Spot the difference	p36
Methodical thinking	p38
49 is the magic number	p40
Testing times	p42
It all adds up	p44
Alphabet soup	p46
Factoring in pairs	p48
Equal shares	p50
The missing link	p52
All or nothing	p54
Number chains	p56
Bite size pieces	p58
Telephone numbers	p60
ISBN numbers	p62

Learning objectives:

- Recognise and extend number sequences formed by counting in steps of constant size
- Recognise multiples of 2, 5 and 10
- Recognise multiples of 3 and 4
- Recognise multiples of 6, 7, 8, 9
- Find all the pairs of factors of any number up to 100
- Factorise numbers to 100 into prime factors
- Know squares of numbers to 10×10
- Know multiplication facts for the 2, 5 and 10 times tables
- Know multiplication facts for the 3 and 4 times tables
- Know multiplication facts for the 6, 7, 8 and 9 times tables
- Know division facts corresponding to the 2, 5 and 10 × tables
- Know division facts corresponding to the 3 and 4 × tables
- Know division facts corresponding to the 6, 7, 8 and 9 × tables
- Use doubling or halving
- Use known number facts to carry out simple multiplications and divisions
- Use the relationship between multiplication and division
- Use closely related facts
- Partition (eg. $47 \times 6 = (40 \times 6) + (7 \times 6)$)
- Understand multiplication as repeated addition
- Understand the operations of multiplication and division and their relationship to each other
- Understand and use the principle of the commutative law
- Develop calculator skills and use a calculator effectively
- Explain methods and reasoning
- Solve mathematical problems and puzzles
- Recognise and explain patterns and relationships
- Generalise and predict

Times Table Tactics

Short activities

Step on

purpose	learning objective	equipment
⊠ exploring the multiplication grid	⊠ recognise multiples of 2, 3, 4, 5, 6, 7, 8, 9 and 10	⊠ OHT of Resource sheet 1 or 2 ⊠ Resource sheet 1 or 2 for each pair ⊠ 1–6 or 1–10 dice or spinner for each pair (see Resource sheet 5)

This activity can be used to introduce the multiplication grid.

pairs

Put Resource sheet 1 or 2 on the OHP and ask children to tell their partners at least ten things about it.

class

Ask individuals to tell you one thing about the grid that their partner told them. Remind them not to repeat something that has already been said.

Encourage children to listen carefully to what others say.

I wonder how you could use the grid to count forwards in 2s? In 4s? In 9s?

Point to the fourth row of multiples and read them out. Tell the children that the numbers in this sequence (4, 8, 12, 16...) are called the multiples of 4. Ask where else they can see multiples of 4.

Multiples of 4 also appear in the fourth column.

pairs

Children take turns to roll the dice or spin a spinner and read out the number. This number indicates which multiple their partner must count in. The dice-roller alone has a copy of the grid. Their partner scores a point for each multiple they say correctly.

Multiple choice

purpose	learning objective	equipment
⊠ practising using a multiplication grid to learn multiples	⊠ recognise multiples of 2, 3, 4, 5, 6, 7, 8, 9 and 10	⊠ OHT of Resource sheet 1 or 2 ⊠ Resource sheet 1 or 2 for each pair ⊠ individual whiteboards (or arrow or digit cards) ⊠ 1–5, 1–6 or 1–10 spinner or dice for each pair (see Resource sheet 5)

class

Put Resource sheet 1 or 2 on the OHP. Point to the numbers in the fourth row of multiples. Remind children that the numbers in the sequence 4, 8, 12, 16 are called the multiples of 4. The first multiple of 4 is 4, the second multiple is 8 and so on.

Make sure the children have understood the language and terminology.

Can you tell me how to find the fourth multiple of 3?

I wonder whether the fifth multiple of 2 is 11?

Turn the OHP off. Spin a spinner twice: the first number represents a column of multiples and the second represents the multiple you want. So if you get 3 and 5, you want the fifth multiple in the column of 3s. Children write the answer on their individual whiteboards (or use arrow or digit cards) and hold these up to face you.

pairs

Each pair needs a spinner or dice and a copy of Resource sheet 1 or 2. One child spins the spinner twice and asks for the appropriate multiple. The other child scores two points for a correct answer found without using the multiplication grid and one point for a correct answer found with the grid. After ten turns, they swap roles.

Children don't need to record their spinner numbers; encourage them to get the answers as quickly as they can.

Hands and fingers

purpose	learning objective	equipment
developing models for multiplication from the imagery of a group of objects	know multiplication facts for the 5 times table (or other multiples)	1–6 or 1–10 dice or spinners (see Resource sheet 5) OHT of Resource sheet 1

class
Roll a dice or spin the spinner, read out the number, and ask children to discuss with their partner how many fingers there would be on that number of hands. Agree the answer and record it on the board. Repeat the process once more.

pairs or groups
Children write down the headings. They roll a dice or spin the spinner and record that number in the 'Hands column'. They then work out the corresponding number of fingers and record that quantity. They do this several times.

Hands	Fingers
3	15
1	5
7	35

class
Collect a set of results by asking each pair for an answer that is not already on the board. Record these in the order they are given.

Put Resource sheet 1 on the OHP with just the numbers in the first and fifth column of multiples exposed.

I wonder how many fingers there are on five hands?

If there are 25 fingers, how do I work out how many hands that is?

Repeat similar questions with the OHP turned off.

To practise other multiples, just choose a suitable context. For example, use eggs and egg boxes for multiples of 6, or cars and wheels for multiples of 4.

This shows the 'hands' and 'fingers' results in order.

This second question hints at division.

Purple patches

purpose	learning objective	equipment
describing relationships between pairs of numbers	recognise and explain patterns and relationships	none needed

class
Write on the board two columns of numbers, one in red, the other in blue. Explain that these columns continue indefinitely.

I wonder how these numbers will continue?

What do you notice about the two sets of numbers?

Can anyone see any patterns or relationships between them?

Red	Blue
1	4
2	8
3	12
4	16
5	20

Accept all contributions and ideas, but ask the children to check them for accuracy. Then ask more searching questions, such as:

How do you know whether 57 will be a blue number?

When 15 is in the red column, how would you work out the corresponding blue number?

How would you work out the corresponding red number for blue 80?

Can you tell me how to work out the last digit of the tenth blue number?

You could use, say, multiples of 3 or 9 as blue numbers instead. Of course, if you do, the relationships described below will differ.

Ideas may include:
- the blue numbers are all even
- $r \times 4 = b$
- $b \div 4 = r$
- $r + b =$ the multiples of 5
- $b - r =$ the multiples of 3

Kangaroo jumps

purpose	learning objective	equipment
developing the imagery of 'jumping forwards' as a model for multiplication	know multiplication facts for the 3 times table (or other tables)	demonstration 0–100 number line copies of Resource sheet 4 OHT of Resource sheet 1

class

Put a counter on zero on the demonstration number line (or use a marker pen) and make jumps of 3 to 3, then 6, 9, 12, 15 and 18.

Can anyone describe what I have just done?

Record the landing places on the board. The children predict the next few numbers you will land on, and explain their reasoning.

Do you think I will land on 30? How can you be sure about this?

How can you work out what numbers I definitely won't land on?

individuals or pairs

Children use Resource sheet 4 to make jumps of 3 from zero. They circle the numbers the counter lands on.

3
6
9
12
15
18

class

Discuss the patterns that emerged on the children's grids.

Put Resource sheet 1 on the OHP and focus on the row of the 3 times table. Practise reciting the multiples of 3 forwards and backwards, starting at different points. Repeat the process with the OHP off.

Point out that the digits of multiples of 3 will always total 3, 6 or 9:

12: $1 + 2 = 3$
15: $1 + 5 = 6$
18: $1 + 8 = 9$

and so on.

Wacky windows

purpose	learning objective	equipment
exploring the idea of common multiples	find simple common multiples	OHTs of Resource sheets 3, 9 and 10 Resource sheet 3 for each pair

class

Display the OHT of Resource sheet 3. Show the children Resource sheet 9. Place it on top of the grid so only the multiples of 6 are showing.

Can anyone say what is special about these numbers?

Replace Resource sheet 9 with Resource sheet 10.

Can you work out which multiples are on display now?

Check whether anyone can say which multiples in this 4s window also appeared in the 6s window. Record the numbers they can remember on the board.

Place Resource sheet 9 on top of Resource sheet 10.

Can you decide what type of numbers these are?

pairs and quartets

Pairs share copies of Resource sheet 3 and each child chooses a different number below 10. One child puts a circle around all of their multiples, and the other puts a square around theirs. They write out the numbers with both a ring and square around them, and swap lists with another pair, who then try to work out the two numbers chosen.

Don't mention multiples of 6 until the children do!

The numbers which show through both 6s and 4s windows are: 12, 24, 36, 48, 60, 72, 84 and 96. Establish that they are all multiples of 12.

Charting the waves

purpose	learning objective	equipment
placing multiples on a chart, and finding which numbers are not multiples	recognise multiples of 2, 3, 4, 5, 6, 7, 8, 9 and 10	Resource sheets 11 and 12 for each pair OHT of Resource sheets 11 and 12 small counters or 1cm cubes 0–9 dice or spinner for each pair (see Resource sheet 5)

class

Display the OHT of Resource sheet 11. Ask the children what they notice about it.

What number patterns can you see and how would the rows continue?

Any idea why 18 appears four times?

If there were more rows, when do you think 24 would appear next?

Turn the OHP off, cover about ten of the numbers with small counters or 1 cm cubes, then turn the OHP on again. Ask which numbers have been covered.

pairs

Each pair has a copy of Resource sheets 11 and 12. They roll a dice, or spin a spinner, twice, and arrange the digits to make a two-digit number. They write this number in the grid on Resource sheet 12 where they think it belongs. If they think it cannot be placed, or belongs on Resource sheet 11, they write it in the box at the bottom of the Resource sheet.

class

Display the OHT of Resource sheet 12. Reveal one row at a time, asking children to tell you the numbers that go in that row, and then fill them in.

Each row contains the multiples of the number shown on the left-hand side.

Children should write each number in every position that it appears in.

Money matters

purpose	learning objective	equipment
exploring the commutativity of the four operations	understand and use the principle of the commutative law as it applies to multiplication	OHT of Resource sheet 13 (enlarged)

individuals and pairs

Display the OHT of Resource sheet 13 with the answers blanked out. Reveal the questions one at a time. Children work out an answer, and a reason, and discuss their results in pairs.

class

Ask for feedback. Help the children to discuss the situations that illustrate commutativity and the ones that don't.

Seven people paying £9 each is not the same as nine people paying £7 each. So seven lots of 9 is not the same situation as nine lots of 7. The situations are different but the overall result is the same.

Similar conclusions can be made about the addition situations in questions 2 and 4, but children need to understand that this does not apply to subtraction and division.

9 divided by 7 is different from 7 divided by 9. The first answer gives you one and a bit, and the other gives you an answer less than 1.

If you have nine sweets and lose seven, you still have two left. But if you have seven sweets you can't lose nine. On the number line, 9 go back 7 takes you to 2. But 7 go back 9 takes you to negative 2.

The answers are given for your reference only.

Don't expect children to talk about this in an abstract way, or to use the word 'commutativity'.

Looking carefully at the rectangular array in question 3 will justify this comment.

Ask the children to illustrate the two division or the two subtraction situations, using counters, on number lines, or by sketching 'real life' situations. This can help bring home the differences to them.

Short activities

And after the break...

purpose
- learning to recognise the multiples of 4 (or another number)

learning objective
- know multiplication facts for the 4 times table (or other tables)

equipment
- OHP calculator
- calculator for each pair
- OHT of Resource sheet 1

You need to know how the constant function works on your calculator so that you can generate multiples of any number, starting at zero. On many calculators you clear the screen to zero, put in the function you want, such as 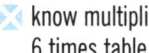, then keep pressing the button.

Make sure that children know the procedure for generating multiples on their calculators, as well as how to get back to zero.

class
Without saying what you are doing, use the OHP calculator to generate multiples of 4, from zero to 24. Encourage children to say what they think is happening.

Now ask if they can remember, in order, the numbers displayed. Record their ideas on the board and check them with the OHP calculator. Return to 24 and generate further multiples. Before each one, ask children to predict what comes next, then check by pressing the equals button. Repeat this process twice more, starting at zero each time but stopping on different multiples of 4.

pairs
Children set up their calculator to generate multiples of 4. One child presses the equals button a number of times (less than ten). Their partner predicts the next multiple. They return to zero and swap roles.

class
Display the OHT of Resource sheet 1 and look at the fourth column of multiples.

What comes after 16? And after 24?

What comes before 36? And before 12?

Turn off the OHP and ask similar questions.

A pressing matter

purpose
- using a calculator to learn the multiples of 6 (or another number)

learning objectives
- know multiplication facts for the 6 times table (or other tables)
- use the relationship between multiplication and division

equipment
- OHP calculator
- calculator for each pair
- OHT of Resource sheet 1

Because this is an adaptation of the activity 'And after the break', children should have done that activity before doing this one.

Establish that the seventh multiple of 6 is 42, the second multiple of 6 is 12, and so on.

This activity touches on division as well as multiplication.

Extension: explore more than ten but fewer than twenty presses.

class
Set the OHP calculator to generate multiples of 6, and tell the children what you have done. Secretly press the equals button seven times then show the number (42) on the display. Ask how many presses you made. Repeat the process twice more and record the results:

multiple	presses
42	7
12	2
30	5

pairs
Pairs set their calculator to generate multiples of 6. They take turns to press the equals button up to ten times, in secret, and hand the calculator to their partner, who guesses how many presses they made.

class
Put Resource sheet 1 on the OHP and focus on the sixth column of multiples.

Any ideas what the fourth multiple of 6 is?

How can you work out which multiple of 6 is 54?

Now turn off the OHP and ask similar questions.

It's a mystery

purpose	learning objectives	equipment
learning to recognise multiples of 2, 4 and 8 (or other multiples)	recognise multiples of 2, 3, 4, 5, 6, 7, 8, 9 and 10 use the relationship between multiplication and division	OHP calculator calculator for each pair OHT of Resource sheet 1

class

Set the OHP calculator to generate multiples of a mystery number (for example, 8). Show zero in the display, then cover it up. While the children watch, press the equals button four times. Then display the result.

How do you work out what the mystery number is? How many presses was that?

mystery number	presses	result
?	4	32
?	10	50
?	7	35
?	3	18

This is an extension of 'A pressing matter'. Children should have done that activity before doing this one.

This will give you the fourth multiple.

Return to zero and repeat the process using another number and different numbers of presses.

pairs

Write up three numbers (say, 2, 4 and 8). One child chooses one of these in secret and repeats what you have done. The other child works out the mystery number.

They play alternately for ten turns, scoring a point for every correct answer.

The range of numbers you use will depend upon the year group. Encourage children to record their work in three columns as you did.

class

Put Resource sheet 1 on the OHP. After two minutes, turn it off.

Who can remember which number 48 is the sixth multiple of?
Any suggestions as to what I should multiply by 5 to make 45?

These questions touch on division, as well as multiplication.

Calculating questions

purpose	learning objective	equipment
finding missing numbers in multiplication equations	know multiplication facts for all the times tables	Resource sheet 1 or 2 for each pair

class

Write four multiplication equations on the board – but for each equation replace one of the digits with a question mark. Choose multiplications to suit the age of the children.

With the children, establish what the missing digits are, using the multiplication grid if necessary.

What do you multiply by 7 to make 35?
What should I multiply 9 by to make 36?

$? \times 7 = 35$
$9 \times ? = 36$
$2 \times 8 = ?6$
$3 \times 7 = 2?$

You need to do four equations because there are four positions for the question mark.

If appropriate, encourage children to stick to tables they are currently practising.

pairs

Each child writes down ten multiplication statements with one digit missing and gives them to their partner to solve.

Make Resource sheet 1 or 2 available as appropriate.

Early finishers could write out more equations for their partner to solve, with two missing elements.

$5 \times ? = ?0$
$6 \times 3 = ??$

When using two elements, there could be more than one solution. For example, the equation
$5 \times ? = ?0$
could be rewritten as:
$5 \times 2 = 10$
$5 \times 4 = 20$ and so on.

Short activities **17**

Box clever

purpose	learning objective	equipment
exploring a multiplication puzzle	use doubling, starting from known facts	OHT of Resource sheet 14 Resource sheet 14 for each pair

To keep the multiplications simple use the numbers 2, 3, 4 and 5 (using 6, 7, 8 and 9, for example, would result in some tricky multiplications).

This gives practice in multiplication strategies such as the use of doubling:
$15 \times 2 = 30$
$15 \times 4 = 60$
$15 \times 8 = 120$

The answers will always be the same, because what they are doing is multiplying all four numbers together:
$4 \times 2 \times 5 \times 3 = 120$

Extensions: use any digits and leave three boxes empty or use multiples of 10 instead of single-digit numbers.

class

Display the OHT of Resource sheet 14. Ask for four different single-digit numbers. Enter these in the top left-hand squares in any order.

With the help of the class, complete the three multiplication equations which go down and the three which go across.

Would it make a difference if the original four numbers were swapped around?

Let the children check this for themselves.

pairs or quartets

Individuals or pairs have grids from Resource sheet 14. They choose four digits to place in the squares. They complete the grid, then copy the array, leaving two of the boxes empty. They swap these incomplete grids with another child or pair to finish.

Chosen numbers

purpose	learning objective	equipment
exploring which two-digit numbers appear in the multiplication grid	recognise multiples of 2, 3, 4, 5, 6, 7, 8, 9 and 10	OHT of Resource sheet 1 0–9 number cards for each child

In these two minutes, children try to remember as many of the numbers on the OHT as possible.

With 4 as the chosen number, 06, 28 and 40 appear in the grid.
$0 + 2 + 4 = 6$
This scores 6 points.

Extension: ask children to find the numbers that produce the maximum and minimum scores.

class

Display the OHT of Resource sheet 1 for two minutes, then turn it off.

Write numbers 0 to 9 in a column on the board. A child chooses one of the numbers (say 4); you form another column of numbers 0 to 9, with zero directly beside this chosen number. Read out the resulting two-digit numbers (6, 17, 28, and so on).

Which of these numbers appear in the multiplication grid? Does 17? Does 28?

Can you say which columns on the grid 6 appears in? And what about 40?

Turn on the OHP again to check which numbers appear in the grid. Give a score to the arrangement by adding together the tens digits of those numbers that do.

Repeat the above process with a different 'chosen number'.

pairs

Play as a game. One child arranges the cards in a column, chooses a number, and forms a second column, then reads out the resulting two-digit numbers. The tens digits of the numbers that appear in the multiplication grid score points, as in the class game, for the second child. Children then swap roles. The child with the most points after five turns each is the winner.

Up and down

purpose	learning objective	equipment
applying knowledge of multiplication facts	know multiplication facts for all the times tables	copies of Resource sheet 1

class

Ask someone to give you two digits between 1 and 9 (the same, or different). Ask someone else to multiply them. Write the relevant equation on the board.

Take the two digits, increase the first by 1, and decrease the second by 1 to form a new multiplication. Establish the difference between the original and the new answer.

> $2 \times 9 = 18$
> $3 \times 8 = 24$
> difference of 6
> (6 more)

Do you think the difference will always be 6? Let's look at some other numbers.

Repeat the process twice more with other digits. Discuss the outcomes.

pairs or individuals

Make Resource sheet 1 available. Set the following investigations for children to choose from:

- What is the largest increase you can get in the products? And the smallest? *(8 and 0)*
- What is the largest decrease you can get in the products? And the smallest? *(10 and 0)*
- Find out if you can ever get a difference of zero. *(yes)*
- What would happen if you went 2 up and 2 down? *(maximum increase 14, decrease 22)*

The differences will vary. For example:

$7 \times 7 = 49$
$8 \times 6 = 48$
difference of 1 (1 less)

The closer together the numbers, the smaller the difference will be, and vice versa.

The aim of these questions is for children to go as far as they are able to, practising what they have learned.

Every which way

purpose	learning objective	equipment
increasing familiarity with multiplication equations	know multiplication facts for all the times tables	OHT of Resource sheet 15 squared paper or copy of Resource sheet 6 for each

class

Display this table on the OHT and ask the class to tell you anything they notice about it.

9	7	6	3
3	9	2	7
2	6	1	2
7	3	2	1

Value all contributions, but focus on the fact that the digits in form multiplication equations:

$9 \times 7 = 63$ $6 \times 2 = 12$
$3 \times 9 = 27$ $3 \times 7 = 21$

When you think they have understood the format ask the children to give you some multiplication equations from the second table:

8	8	6	4
3	7	2	1
2	5	1	0
4	6	2	4

Now present the third table and ask what children notice about it:

6	3	1	8
9	4	3	6
3	8	2	4
4	5	2	0

All four rows form correct multiplication equations, but the four columns do not.

pairs or individuals

Each pair outlines some blank 4 × 4 grids on squared paper.
They then fill these in with numbers to try and get the correct multiplication equations in all four rows and in all four columns.

Children may well get all four rows reading correctly but not all the columns. Give credit for any column that reads correctly.

The spying game

purpose
- using multiplication facts to solve problems

learning objective
- solve multiplication problems

equipment
- 0–9 number cards for each child
- Resource sheet 1 or 2 for each pair

For a simpler activity, stick to numbers no higher than 6.

Note that there can be more than one solution to any given code.

Make Resource sheets 1 or 2 available to allow children to look for likely solutions.
For extra support give them one of the single digits you started with. Also ask questions to help them narrow the range of solutions: 'If the code is 19, what does that tell you about the answer to the multiplication? It must be less than 19 and contains an odd number'.

class

Ask someone to choose two digits then ask a different child to multiply them. Write the multiplication equation on the board.

The class add the two single digits and the answer: this total is the special code for that particular multiplication. Write the 'code' by the side of the original multiplication equation. Repeat this process three or four more times.

Now write down the code for a multiplication you have done secretly. Set children the task of finding out what the original multiplication was.

Give children a few minutes to work on this. As a class, check that solutions are correct.

pairs

Each child secretly writes down five different multiplications. They carry out the above process and give their partner the codes. The partner then has to work out the original multiplications, using digit cards if they wish.

6 and 5
$6 \times 5 = 30$
$6 + 5 + 30 = 41$

$6 \times 5 = 30$ (41)
$5 \times 4 = 20$ (29)
$8 \times 8 = 64$ (80)
$2 \times 7 = 14$ (23)

Code: 19
Answer:
4×3 (or 3×4)
or
1×9 (or 9×1)

Scrambled numbers

purpose
- practising multiplication facts by solving puzzles

learning objective
- solve multiplication problems

equipment
- Resource sheet 1 or 2 for each pair

You may want to jot down the originals, in private, as an aid to memory.

Acknowledge that most 'jumbles' have two partner solutions.
4123 could be
$3 \times 4 = 12$ or $4 \times 3 = 12$
But 4614 can only be
$4 \times 4 = 16$

Simplification: only use numbers between 2 and 6.

Extension: include numbers that are greater than 10.
$12 \times 6 = 72$ or $6 \times 12 = 72$
76122

class

Write a multiplication equation on the board and below it write out the digits, in order, but without any signs. Underneath that, write down the same digits but jumbled up. As soon as you have written the jumbled version down, rub out the previous two lines.

Repeat this process three times so that all you have on the board is four sets of jumbled digits. Then point to each set of digits in turn, and ask the children what the original multiplication was.

Can you say what the equation could have been?

Can anyone think of any other possible solutions?

pairs

Each child secretly writes down ten multiplication equations, jumbles up the digits and then gives the jumbled version to their partner to unravel.

Make Resource sheets 1 or 2 available, but encourage children to work without them where possible.

$4 \times 7 = 28$
4728
2748

2748
1638
4123
7396

Hundreds and thousands?

purpose
exploring the different products that appear in the multiplication grid

learning objective
recognise products

equipment
OHT of Resource sheet 1

class
Ask the children to guess how many different products appear in the 10×10 multiplication grid, without looking at it. Tell children that duplicates do not count. For example, 4 appears twice but should only be counted as one product.

Record a selection of the children's estimates. Discuss the different methods that can be used to find the correct number of products in the grid.

estimates
80
50
110
75

pairs or individuals
The children list all the different products they can find.

Early finishers can estimate, then establish, the number of products if the grid were continued up to 12×12 or 15×15.

class
Collect some of the children's answers, and confirm or disprove these by checking for duplicate products on the OHT of Resource sheet 1.

Compare the actual number of products with the estimates.

Are you surprised by how few products there are in the grid?

Which estimate is closest to the actual answer? How close was it?

Up to 10×10 there are 42 different products. To find them, children can count the products, crossing off duplicates. Explain that the diagonal line of square numbers divides the grid into two identical sets of numbers. So the children can discount 45 numbers immediately.

Up to 12×12 there are 59 different products. Up to 15×15 there are 90 different products.

Broken keys

purpose
exploring equivalent operations

learning objective
use a range of mental calculation strategies

equipment
calculators

class
Write this on the board:

$6 \times 7 =$

Ask the children to imagine that the 7 key doesn't work on their calculator, but that you still want them to use their calculators to find the answer.

Can you think of a way to work out the answer without using the 7 key?

After a few minutes, get some feedback, asking for descriptions of the different strategies the children used. Repeat each strategy the children describe, and highlight what is different or similar about them.

pairs or individuals
Children imagine that the 6 key has been broken, rather than the 7 key.

I wonder what we can do now to work out the answer to 6×7?

Early finishers choose another multiplication and see whether the strategies they have already used work with these numbers.

If children know the answer already, tell them to prove that their answer is right.

You may need some ideas up your sleeve to share with the children:

$6 \times 6 + 6 = 42$

$6 \times 10 - 6 - 6 - 6 = 42$

$3 \times 14 = 42$, and so on.

A strategy that works when the 6 key is broken is to press $7 \times 2 \times 3$, that is, split the 6 into its two factors. You cannot use this strategy when the 7 key is broken because 7 is a prime number.

Short activities **21**

Times Table Tactics

Long activities

Checks and stripes

purpose
- becoming familiar with the multiples of 3 (or another number)

learning objectives
- recognise and extend number sequences formed by counting in steps of constant size
- recognise multiples of 3

equipment
- OHTs of Resource sheets 3 and 16
- Resource sheet 16 or 17 for each pair

This activity explores the patterns of multiples on a 1–100 grid and other grids, building on the short activity 'Kangaroo jumps' (page 14). It will help children develop the idea that multiples occur in a regular pattern. You can do this activity equally well with multiples of 4, 6 or another digit, adjusting the content of the plenary by choosing the grid that will produce multiples in two columns.

pairs or groups

Present the OHT of Resource sheet 3 and ask the children to tell each other about any patterns they can see.

1	2	3	4	5	6	7	8	9	10
11	12	13	14	15	16	17	18	19	20
21	22	23	24	25	26	27	28	29	30
31	32	33	34	35	36	37	38	39	40
41	42	43	44	45	46	47	48	49	50
51	52	53	54	55	56	57	58	59	60
61	62	63	64	65	66	67	68	69	70
71	72	73	74	75	76	77	78	79	80
81	82	83	84	85	86	87	88	89	90
91	92	93	94	95	96	97	98	99	100

The aim here is to lay the foundations for pattern spotting later in the lesson.

The short activity 'Kangaroo jumps' uses a 0–99 grid, as the idea is to jump in 3s from zero, mimicking the jumps done on a number line. This activity uses a 1–100 grid to explore the patterns of multiples more deeply.

class

Ask for and explore children's ideas, accepting all contributions. Follow up ideas, asking for reasons and justifications.

I wonder why the ones digits are the same in each column?

Can anyone tell me why the tens digits increase by 1 as you move down a row?

Circle the first six multiples of 3 (up to 18) on the grid.

1	2	3	4	5	6	7	8	9	10
11	12	13	14	15	16	17	18	19	20
21	22	23	24	25	26	27	28	29	30
31	32	33	34	35	36	37	38	39	40

Ask children what you have done. Establish that the circled numbers are called the multiples of 3.

Some children may say you have missed out two numbers or you are jumping in 3s. Acknowledge these ideas but focus on the word 'multiple'.

Which five numbers should I circle next to continue the sequence?

Discuss with the children whether they prefer to use the visual or numerical pattern to continue the sequence.

When you have circled all the multiples of 3, ask the children to put into words any visual patterns and any number patterns they can see. In order to help children see the number patterns, write out the multiples so that those with the same ones digits are under each other.

3	6	9	12	15	18	21	24	27	30
33	36	39	42	45	48	51	54	57	60

Times Table Tactics

pairs

Each pair has a number grid from Resource sheet 16 or 17 and circles all the multiples of 3, then writes a statement about the visual pattern. For example:

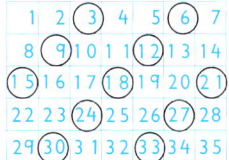

"The multiples of 3 make diagonal lines on the grid."

Early finishers can:

- look at multiples of 4 on the same grid
- work with another grid
- share and discuss their results with another pair or group
- design a grid in which a multiple of their own choosing would produce a particular pattern (say, diagonal or vertical)

plenary

Ask children to describe their grid and their patterns. Discuss why different grids produce different patterns.

Display the grid of 6s to 102, which has the multiples of 3 in two columns.

What kind of grid would have the multiples of 3 in one vertical column?

Display any of the grids, and circle 15.

Can you remember the multiple of 3 that comes before 15?

What is the multiple of 3 after 15? And the multiple two on from that?

Continue until you have circled all the multiples of 3. Cover the grid up. Have the class practise reciting the multiples of 3 forwards and backwards, starting from various points in the sequence.

This would be a grid in 3s:

1	2	3
4	5	6
7	8	9

Although children do not need to learn multiples of 3 above 30 by heart, they should become familiar with them. You could also explore how, in multiples of 3, the digits always add up to a multiple of 3:

12: $1 + 2 = 3$

18: $1 + 8 = 9$

24: $2 + 4 = 6$

63: $6 + 3 = 9$

and 99: $9 + 9 = 18$

Something for the children to work on at home

Design your own number grid and circle all the multiples of 3. Or take a sheet of squared paper and make a spiral, zigzag or triangle grid, and circle all the multiples of 3.

Figure out the fours

purpose	learning objectives	equipment
identifying and working on facts from the 4 times table	recognise multiples of 2, 3, 4, 5, 6, 7, 8 and 9 know multiplication facts for the 2 to 9 times tables use known number facts to carry out simple multiplications and divisions	OHT of Resource sheet 18 1–10 number cards for each child doing activity 1 Resource sheet 1 for each pair, and 7 if doing activity 2 0–9 dice or spinner (see Resource sheet 5)

This activity helps children learn the facts in the 4 times table. However, you can adapt it for any times table from 3 to 9 by using different resources, based on those given here for the 4 times table.

class

Display an OHT of Resource sheet 18 and write 4 at the top. At random, point to a multiplication in the left-hand column and ask for the result of operating on 4 with that multiplication.

The children call out the answer in unison. Don't write in any answers. If there are products they don't know, tell them the answer. Repeat and rehearse as many times as you think necessary.

Or use any other number you want to practise.

At this stage don't worry if they don't know all the answers. By asking the children to call the answers out, but not write them down, you are encouraging children to try to remember them.

Don't place too much importance on the answers in the right-hand column. The purpose is to make use of – and so practise – the products up to 4 × 10.

Now point to any multiplication in the right-hand column and ask for the result of multiplying 4 by that number. If the children don't know the answer, ask which answer or answers in the left-hand column would offer a clue.

How could you find the answer to 4 multiplied by 13?
Can you use 4 multiplied by 10 to help you? Or switch the numbers round and do 13 multiplied by 4?

To do 4 × 20, children may double the answer to 4 × 10. Or for 4 × 12 they may add the result of ten 4s and of two 4s. Accept or suggest appropriate strategies for each multiplication and fill in the answers for the right-hand column.

4 × 11 = 44
4 × 12 = 48
4 × 13 = 52
4 × 14 = 56
4 × 15 = 60
4 × 16 = 64
4 × 17 = 68
4 × 18 = 72
4 × 19 = 76
4 × 20 = 80

pairs

Pairs do one of the following two activities. Get early finishers to watch as another pair do the activity they haven't yet done. They can learn from them what to do, and then do it for themselves.

If children know the 4 times table off by heart they may still need to do these activities because they may not know the facts out of sequence. If they do know the facts out of sequence, give them another times table to work with.

activity 1

Child A has Resource sheet 1. Child B has 1–10 number cards, shuffled and placed in a pile. Child B turns up a card and either multiplies 4 by the number shown, or multiplies the number shown by 4 (children should know that these operations give the same result). Child A checks the answer using the multiplication grid. If correct, the card is put to one side. If incorrect, Child A announces the correct answer and puts the card to the bottom of the pile. They keep going until Child B has got all the multiplications correct – that is, all cards have been discarded. They then swap roles.

"Four times seven is 28" or "Seven 4s equals 28".

activity 2

Child A has Resource sheet 1. Child B has cards cut from Resource sheet 7 showing multiples of 4 up to 40, shuffled and placed in a pile. Child B turns up a card and says the multiplication that corresponds to the multiple on the card.

"I've got 20. That's four multiplied by five" or "I've got 20. That's four 5s".

As before, Child A checks the answer on the grid and, if correct, puts the card to one side. If incorrect, Child A announces the correct answer and puts the card to the bottom of the pile. They keep going until Child B has got all the multiplications correct – that is, all cards have been discarded. They then swap roles.

plenary

Write the numbers 1 to 10 on the board. Ask one child to pick any one of those numbers. Ask another child to double that number. And another child to double that answer. On the board, record the number chosen and, by its side, the result of the double double. For example:

5	20
9	36
7	28

Repeat. Keep going until all ten numbers have been used. Ask the class what they notice (numbers in the right-hand column are 4 times the numbers in the left-hand column).

Now ask the class a number of quick-fire questions, such as:

What is 5 multiplied by 4? 7 multiplied by 4? 6 multiplied by 4?

Rub out the second column and ask a similar set of questions.

> ### Something for the children to work on at home
>
> Write a list of the multiples of 4. Take each number in turn and multiply each of the digits by 4, and then add the results. Explore what kind of numbers you get.
>
> $4 \rightarrow 4 \times 4 = 16$
> $8 \rightarrow 8 \times 4 = 32$
> $12 \rightarrow (1 \times 4) + (2 \times 4) = 12$
> $16 \rightarrow (1 \times 4) + (6 \times 4) = 28$, and so on

The results should all be multiples of 4.

Explain that brackets are used to 'hold together' the multiplications.

Seven days

purpose
- exploring repeated addition expressed as a multiplication

learning objectives
- understand multiplication as repeated addition
- know multiplication facts for the 7 times table
- develop calculator skills and use a calculator effectively

equipment
- OHP calculator
- calculator for each child
- 1–10 dice or spinner for each pair (see Resource sheet 5)

Children need to understand that multiplication is, in effect, repeated addition. This activity reminds children of this, and introduces them gently to the 7 times table, with the use of a calculator.

class

How many days there are in one week? What about three weeks?

Ask children how they worked out the answer. Accept all ideas but highlight the method of totalling three 7s.

One of the easiest ways to find out how many days there are in three weeks is to add three 7s together.

Ask how many 7s must be added together to find out the number of days in eight weeks. Establish the answer and write this on the board:

$7 + 7 + 7 + 7 + 7 + 7 + 7 + 7 =$

Now ask children to find the total using their calculators. When they have got the answer, ask them to work out how many key presses they had to make.

Does anyone know a quicker way of using their calculator to add eight 7s? Can you use fewer key presses?

Use your OHP calculator to show that
$7 \times 8 =$
gives the same result as
$7 + 7 + 7 + 7 + 7 + 7 + 7 + 7 =$

Explain that when $7 \times 8 =$ is entered, the calculator very quickly adds together eight 7s.

Now ask the children to use their calculators to find out how many days there are in fifteen weeks.

pairs

Pairs need a 1–10 dice or spinner, a calculator, and a piece of paper with two headings, 'Weeks' and 'Days'. They roll the dice and enter that number in the 'Weeks' column. They work out how many days there are in that number of weeks, and record the answer in the 'Days' column. They then cover up both numbers with a piece of paper. They repeat this at least twenty times.

Weeks	Days

If you think that children may know that the answer is 56, use thirteen weeks instead of eight.

Adding eight 7s requires sixteen key presses. Some children may have pressed $8 \times 7 =$ or $7 \times 8 =$. If so, ask them to explain what they did.

You might wish to explain that 7×8 can be read as '7, eight times', '7 multiplied by 8' or 'eight 7s', and that they all really mean 'eight 7s added together'.

You need to make sure that they have done 7×15 or 15×7 rather than keyed in 7 fifteen times!

Encourage children to use the calculator only when they need to.

The same number may be thrown several times; the purpose of covering up previous answers is to challenge children to remember them.

Provide those children who finish early with a magic square where each row, column and diagonal adds up to 15.

8	1	6
3	5	7
4	9	2

Ask them to multiply each number by 7 and see if it is still a magic square — that is, if each row, column and diagonal adds up to the same total.

plenary

Write up a partly completed table, like this:

Point to any one of the numbers in the 'Weeks' column and ask someone to tell you how many days there will be in that number of weeks. Write in their response (even if incorrect). Continue until you have completed the table.

Children may challenge incorrect responses. If not, wait until the end and go through the answers in order of size, checking for inaccuracies.

We've got 7... 14... 22... is that right?

Cover up the whole table and ask questions like:

Weeks	Days
5	
3	
7	
1	
9	
2	
4	
10	
6	
8	

Can you work out how many days there are in four weeks?

We know what 7 plus 7 plus 7 equals, so what will 3 multiplied by 7 equal?

Who can remember what four lots of 7 are?

I wonder how many 7s make 28?

If the number of weeks was put in order, children would be able to add or subtract 7 from adjacent answers. Jumbling them up requires children to remember the corresponding links.

Vary the language you use.

Something for the children to work on at home

- Write out the multiples of 7 from 7 to 70 in a column. Now copy out the ones digits only and look from the bottom of this list to the top.
- Some other multiplication tables have the same ones digits but in a different order. Can you work out which times tables have the same ones digits as the 7 times table?
- Are there any other multiplication tables that have the same ones digits as each other?

The 1, 3 and 9 times tables have the same ones digits as the 7 times table.

The 2, 4 and 6 times tables share the same ones digits.

Long activities

Double this, halve that

purpose	learning objectives	equipment
using doubling and halving to find unknown multiplication facts	use doubling and halving, starting from known facts explain methods and reasoning recognise and explain patterns and relationships generalise and predict	copies of Resource sheet 1 (optional) calculators for each child

If children forget multiplication facts, they need to have ways to figure them out mentally. Doubling and halving are valuable strategies; they are explored in this activity.

class

Write a multiplication equation on the board.

$5 \times 4 = 20$

Ask someone to double the number being multiplied and rewrite the equation replacing that number with its double. Establish the product and complete the equation.

$5 \times 4 = 20$
$10 \times 4 = 40$

Ask someone else to rewrite the original equation underneath the other two — this time with the multiplier doubled. Establish the product and complete the equation.

$5 \times 4 = 20$
$10 \times 4 = 40$
$5 \times 8 = 40$

Repeat the process with a different starting equation.

The product is doubled.

Can you tell me what happens to the product of two numbers when you double either one of those numbers?

Repeat the process with a new starting equation but this time halve each number in turn.

$4 \times 8 = 32$
$2 \times 8 = 16$
$4 \times 4 = 16$

The product is halved.

What seems to happen to the product of two numbers when you halve either one of those numbers?

pairs or individuals

Children investigate what happens to the product of two even numbers when:

- you double both numbers being multiplied
- you halve both numbers being multiplied
- you double the first one and halve the second
- you halve the first one and double the second

Children can use a calculator or Resource sheet 1 to help them.

Children should choose one of these strategies to tackle the investigation:

- *choose two numbers to multiply and do all four procedures on that pair of numbers; then repeat this with another pair, and another pair*

- *explore one procedure with several pairs of numbers before moving on to the next procedure*

Children who finish early can try the same investigations with any pair of numbers, odd or even.

plenary

Bring the class together to share and discuss their findings. At the appropriate time, draw out these 'rules':

- when you double both numbers that are being multiplied, their product is quadrupled (made four times as big)
- when you halve both numbers that are being multiplied, their product is quartered (so it is worth one fourth, or one quarter, of the original)
- when you double one of the two numbers being multiplied, and halve the other, their product stays the same (does not change or has no effect)

$2 \times 2 = 4$	$4 \times 4 = 16$
$4 \times 6 = 24$	$2 \times 3 = 6$
$8 \times 6 = 48$	$4 \times 12 = 48$

Write these column headings on the board.

I know this	So I can work out this	How?

In the first column write a multiplication equation, and in the second write a related but incomplete equation, whose answer can be found by doubling or halving. Tell the children that you don't want the answer to each equation but the strategy they would use. The answers can be worked out later.

Draw out the *strategies*, rather than the correct answers.

I know this	So I can work out this	How?
$7 \times 10 = 70$	$7 \times 5 = ?$	By halving 70
$3 \times 7 = 21$	$6 \times 7 = ?$	By doubling 21
$4 \times 4 = 16$	$8 \times 8 = ?$	By doubling 16 twice

Something for the children to work on at home

Give the children these questions:

Someone says that to multiply any number by 4, you simply double its double. Is this true?

Someone says that to multiply a number by 25, you simply multiply it by 100 and then halve the answer. Is this true?

Yes.

No: quarter the answer.

Long activities **31**

Building blocks

purpose	learning objective	equipment
analysing ways of breaking multiplications down into more manageable ones	use doubling or halving; use closely related facts	OHT of Resource sheet 19 squared paper or Resource sheet 6 for each child OHT of Resource sheet 6 (optional)

In this activity, children use doubles, multiples of 10 and multiples of 5 to establish facts that they cannot instantly recall.

class

Present an OHT of Resource sheet 19 or give copies to children to complete.

	1	2	3	4	5	6	7	8	9	10
× 1										
× 2										
× 10										
× 5										

With the children, agree the number that belongs in each cell and complete the grid.

	1	2	3	4	5	6	7	8	9	10
× 1	1	2	3	4	5	6	7	8	9	10
× 2	2	4	6	8	10	12	14	16	18	20
× 10	10	20	30	40	50	60	70	80	90	100
× 5	5	10	15	20	25	30	35	40	45	50

pairs

In pairs, ask the children how to use this information to work out the answer to 8 multiplied by 3. Even if they know the answer, they must try to show how it can be worked out using this information and this information only.

Acknowledge any ideas that work. These may include:

- *splitting 8 × 3 into 8 × 1 plus 8 × 2*
- *thinking of 8 × 3 as eight 3s, then starting with ten 3s, and subtracting two 3s*

class

After an appropriate amount of time, bring the class together and discuss their ideas.

Why does 8 multiplied by 1 added to 8 multiplied by 2 give the same answer as 8 multiplied by 3?

Sketch a grid of squares on the board (or use an OHT of Resource sheet 6) and outline 8 × 1 and below that, 8 × 2. Then outline a 8 × 3 rectangle and demonstrate how the numbers of squares are the same:

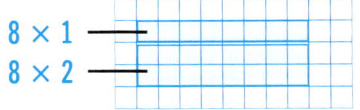

Write this equation and explain how it describes the situation:

$(8 \times 1) + (8 \times 2) = 8 \times 3$

Show that brackets are used to 'hold together' the two multiplications.

pairs

Write on the board 7×6 and 6×7. The children again use the information on the grid to work out the answer to either multiplication, and test this out for themselves using squared paper, or Resource sheet 6.

Children can choose other pairs of numbers to investigate in a similar way.

plenary

Look with the children at the original table, and ask questions such as:

If you know that 6 multiplied by 2 is 12, how could you work out 6 multiplied by 4?

$6 \times 2 = 12$
$6 \times 4 = 24$

If you know that 3 multiplied by 10 is 30, and 3 multiplied by 1 is 3, what else do you know?

$3 \times 10 = 30$
$3 \times 1 = 3$
$3 \times 11 = 33$ and $3 \times 9 = 27$

If you know that 7 multiplied by 10 is 70, and 7 multiplied by 8 is 56, what else do you know? Could you use this information to find 7 multiplied by 28?

$7 \times 10 = 70$
$7 \times 8 = 56$
$7 \times 18 = 70 + 56 = 126$
$7 \times 28 = 70 + 70 + 56 = 196$

Early finishers can explore using subtraction to get the answer:

$(7 \times 10) - (7 \times 4) = 7 \times 6$

Write these equations on the board as you say them.

Sketch rectangles on the OHT grid to support these deductions.

$3 \times 11 = 33 \begin{cases} 3 \times 10 = 30 \\ 3 \times 1 = 3 \end{cases}$

$7 \times 18 = 126 \begin{cases} 7 \times 10 = 70 \\ 7 \times 8 = 56 \end{cases}$

Something for the children to work on at home

Start with $8 \times 2 = 16$, and write down all the other multiplications you can derive from this fact.

Heading into the great unknown

purpose
- developing strategies for working out unknown products

learning objectives
- know multiplication facts for the 8 times table (or another table)
- use known number facts to carry out simple multiplications; use closely related facts

equipment
- calculator for each pair

Working with a multiplication table whose facts children don't need to know, such as the 25 times table, allows them to focus on strategies for working out unknown facts rather than simply trying to memorise them. They can then use these strategies to work on the 8 times table (or any other table).

class

Write a jumbled list of the 25 times table, without answers, on an OHT. Briefly display the list. Now cover this with a sheet of paper, reveal the first equation and ask a child to read it out. Ask another child to try and work out the answer. If they know it, complete the equation. If they don't, just put a question mark.

$25 \times 2 = 50$ or $25 \times 2 = ?$

$25 \times 2 =$
$25 \times 8 =$
$25 \times 7 =$
$25 \times 5 =$
$25 \times 3 =$
$25 \times 10 =$
$25 \times 6 =$
$25 \times 9 =$
$25 \times 1 =$
$25 \times 4 =$

Repeat the process until all ten equations have been revealed. You will probably end up with just some of the equations completed.

Can you use any of the answers on the board to work out the answers we don't have?

Discuss the strategies the children can use and, with the class, fill in the rest of the answers.

$25 \times 2 = 50$
$25 \times 8 = ?$
$25 \times 7 = ?$
$25 \times 5 = 125$
$25 \times 3 = ?$
$25 \times 10 = 250$
$25 \times 6 = ?$
$25 \times 9 = ?$
$25 \times 1 = 25$
$25 \times 4 = ?$

Look for strategies that build on 'easy' facts. For example:

25×2 is double 25

25×3 is (25×1) plus (25×2)

25×5 is half of 25×10

Making children work with some unfamiliar multiplications ensures that they concentrate on the process rather than the calculation.

Now write this on the board:
Choose any of the missing multiplications (up to 37 × 10) and ask the children how they can use the existing information to obtain the answer.

$37 \times 1 = 37$
$37 \times 2 = 74$
$37 \times 4 = 148$
$37 \times 8 = 296$

Don't say the answer – say how you might use what we have done so far to find answers to the missing multiplications.

Again, look for strategies such as doubling (for × 2, × 4 and × 8), or multiplying by 10 and halving (for × 5).

pairs

Write these multiplications on the board:
Ask each child to explain to their partner how they could use the information given to work out the missing multiplications (up to 8 × 10).

$8 \times 1 = 8$
$8 \times 2 = 16$
$8 \times 4 = 32$
$8 \times 8 = 64$

You could use facts from any of the other times tables instead.

pairs

Child A has a calculator. Child B chooses a number greater than 10 to multiply 8 by, using information from the abridged 8 times table on the board to help them.

I'm going to multiply 8 by 12. It's ten 8s and two 8s, which is 80 and 16, which equals 96.

Child A checks the answer with the calculator. If it's correct, Child B picks a new multiplier and repeats the process; otherwise, they swap roles. If a player is successful five times in a row, they win the game, and the pair start a new game after exchanging roles.

Some children could use number cards, as they do in the activities from 'Figure out the fours' (p26–7).

plenary

Work with the children to write up another table, such as the 15 times table, making use of known facts.

$15 \times 1 = 15$
$15 \times 2 = 30$
$15 \times 3 =$
$15 \times 4 = 60$
$15 \times 5 =$
$15 \times 6 =$
$15 \times 7 =$
$15 \times 8 = 120$
$15 \times 9 =$
$15 \times 10 = 150$

Which shall we start with? What is an easy fact?

We've done 15 multiplied by 2, 15 multiplied by 4, and 15 multiplied by 8, all by doubling. I wonder how we could do 15 multiplied by 9?

Two useful methods of multiplying 15 × 9 in this context are:

$(15 \times 8) + (15 \times 1)$

and

$(15 \times 10) - (15 \times 1)$

Something for the children to work on at home

Use this information:

$9 \times 1 = 9 \qquad 9 \times 3 = 27 \qquad 9 \times 5 = 45$

to find the answers to 9 × 2, 9 × 4, 9 × 6, 9 × 7, 9 × 8, and 9 × 9.

Explain how you worked out each answer.

Spot the difference

purpose	learning objectives	equipment
establishing the principle of commutativity	know multiplication facts for the 6 times table understand and use the commutative law of multiplication develop calculator skills and use a calculator effectively	counters and OHP strips showing multiples of 6 or another number (see Resource sheet 7 or 8) calculator for each child OHT of Resource sheet 20

In this activity children explore the commutativity of multiplication, and use it to help them learn the multiplication facts for the 6 times table. The activity can easily be adapted to a different table.

class and pairs

Present this arrangement of counters on the OHP:

Ask children to describe what they see and anything else they wish to tell you about the situation. Value all contributions.

Say that you want to change this to seven groups of six. Children discuss in pairs whether you will need:

- *fewer counters*
- *the same number of counters*
- *more counters*

Ask individuals what they think and why. Then collect the odd end counters together, so that the arrangement looks like seven groups of six.

Establish that there are six groups of seven counters which can be recorded as 7 × 6 (7 multiplied by 6).

Ask one child to do 7 × 6 = on their calculator and another to do 6 × 7 = . Compare the answers.

Do you think 6 multiplied by 7 and 7 multiplied by 6 are the same?

Remove three of the groups of six, leaving four groups.

How many groups of six counters are there now?

The 'why' is very important. For example, 'you need the same because six 7s is the same as seven 6s' or 'it's the same number, just arranged differently'.

Would I need fewer counters, the same number of counters or more counters, in order to make six groups of four?

Discuss this fully with the children. Establish that you would need the same number, and demonstrate this by moving the counters.

They are not, strictly speaking, the same. Only their products are the same.

When you are trying to remember multiplication facts, you can always turn them around if it helps. So if you can't remember what four 6s are, see if you can do six 4s.

You need the same number. But it has been known for children to think it only works with seven groups of six!

pairs

Pairs need a calculator and a strip of paper with multiples of 6 (you could use other multiples, depending on the children). Child A points to a number and says how many groups of 6 that number of counters would make. Child B then checks the multiplication on their calculator.

Eighteen. That's three groups of 6; do 6 × 3 on the calculator.

If the answer on the calculator is the same as the number that was pointed to, they put a tick by the number. If the answers differ, they put a cross. When they have dealt with all the numbers, they count how many they have ticked. They then reverse roles, forming a second column of ticks and crosses.

Ask the children to make these strips. Alternatively, you could make several on one A4 page and photocopy them, or photocopy Resource sheet 7 or 8.

Remind children that if they don't recognise, say, the number of 6s in 24, they should see if they can remember other tables that 24 appears in.

"Six 4s are 24 so four 6s must be 24 as well."

| 6 |
| 12 |
| 18 |
| 24 |
| 30 |
| 36 |
| 42 |
| 48 |
| 56 |
| 60 |

plenary

Present this on the board or OHP:

| 6 + 6 + 6 + 6 + 6 + 6 + 6 + 6 + 6 + 6 = |
| 10 + 10 + 10 + 10 + 10 + 10 = |

Can you guess which total is larger? Why do you think that?

How could you check this with multiplication rather than addition?

Now display an OHT of Resource Sheet 20. Children discuss in pairs:

- how it shows that the two long additions
  (6 + 6 + 6 + 6 + 6 + 6 + 6 + 6 + 6 + 6
  and 10 + 10 + 10 + 10 + 10 + 10)
  have the same total
- how it shows that the two multiplications 6 × 10 and 10 × 6 have the same answer

Get feedback, and discuss this fully.

Focus the discussion on:
- the number of counters in each row and column
- the fact that 6 × 10 is a quick way of adding ten 6s together, and 10 × 6 is a quick way of adding six 10s together
- the total is 60, whatever the method

Something for the children to work on at home

Write down all the multiples of 6 from 12 to 96, then swap the digits of each multiple to form another set of two-digit numbers (for example, 12 becomes 21 and 36 becomes 63). Now answer the following questions:
- Which numbers in the new set of two-digit numbers are also multiples of 6?
- Which numbers in the new set of two-digit numbers are not multiples of 6? These numbers all belong to one multiplication table. Can you work out which one?

They all belong to the 3 times table.

Methodical thinking

purpose
- analysing a range of multiplication strategies

learning objectives
- use doubling or halving; use closely related facts; partition
- extend number sequences formed by counting in steps of constant size

equipment
- OHT of Resource sheet 21
- Resource sheet 21 for each pair
- OHT of Resource sheet 6, or counters

Children look closely at the recordings of four different methods for working out an unknown multiplication fact, then attempt to follow the same methods for a similar problem. Some of the methods used in this activity are introduced in the activities Building blocks (page 32–3) and Spot the difference (page 36–7).

pairs

Display Resource sheet 21. Pairs discuss the following questions, and try to agree how they would answer them:

Look carefully at each way of working out 8 multiplied by 9. How does each method work?

Imagine you have to explain the method to someone else, a younger brother or sister perhaps. What would you say to them?

class

Bring the class together and help them read out the multiplication equations from the OHT.

8 multiplied by 9 equals 8 multiplied by 5 plus 8 multiplied by 4.

Discuss their ideas about the four methods and illustrate these, if appropriate, with arrays of squares or counters on the OHP.

The first one was done in two bits. First five lots of 8, then four lots of 8.

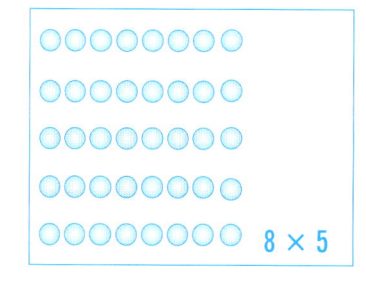

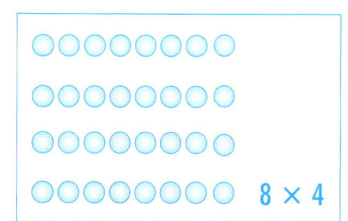

8×5 8×4

8×9

Label each method using language agreed with the children.

Take the time to help children fully understand the structure of each method so that they can apply it to other multiplications.

In more detail, the methods are:

- Splitting the second number into two convenient parts
- Multiplying by 10 and subtracting, because × 10 is an easy multiplication to remember
- Counting in step sizes relating to the multiplier
- Using the fact that multiplication is commutative

Ask if anyone can think of any other ways to multiply 8 by 9. If they can, add these to the list and label them with the child's name.

> A Splitting up
> B Using × 10
> C Counting in 8s
> D Swapping over

Someone has suggested doubling 9 repeatedly.

Eight multiplied by 9 is the same as 9 multiplied by 8.

How could we double 9 repeatedly to get the answer?

$8 \times 9 = 9 \times 8$

$9 \times 2 = 18$

$9 \times 4 = 36$

$9 \times 8 = 72$

pairs

Hand out Resource sheet 21. Children try to use each of the four methods: splitting, using × 10, counting in steps and swapping over, to do 9 × 6. If they are not yet familiar with brackets, they can find their own way to record the calculations – or you can give them a quick lesson in the use of brackets. Whatever they do, ask them to make sure they can read out their recordings.

They should also:
- add any of their own methods to the list
- decide which method they think is the easiest
- decide which they think is the hardest

> There is more than one way to carry out the first method, splitting up. For example, a child might come up with:
>
> $(9 \times 4) + (9 \times 2)$ or
> $(9 \times 5) + (9 \times 1)$

plenary

Bring the class together and share their ideas. Discuss these fully.

What difference would it have made if the problem had been 6 × 9 instead of 9 × 6?

I wonder if it is easier to do 9 × 3 or 3 × 9? Why do you think this?

Can you give another multiplication where 'swapping over' would be useful?

> **Something for the children to work on at home**
> Find as many methods as you can to work out 12 × 3.

49 is the magic number

purpose
- demonstrating strategies for working out multiplication facts up to 10 × 10

learning objective
- know multiplication facts for all times tables to 10 × 10

equipment
- OHT of Resource sheet 1
- OHT of Resource sheet 6 to make a multiplication grid

Once children have learned techniques such as doubling and partitioning, they will find that there is only one multiplication fact they still need to learn: 7 × 7 = 49.

class

Display the OHT of Resource sheet 1. Explain to the children that you are going to demonstrate which of the 100 multiplication facts they need to learn. Replace Resource sheet 1 with an empty multiplication grid (you could use Resource sheet 6 to make one). With the children, work through each multiplier in turn.

×	1	2	3	4	5	6	7	8	9	10
1										
2										
3										
4										
5										
6										
7										
8										
9										
10										

The 1 times table is easy to remember because the numbers stay the same.

With the children's help, fill in the first row and column of multiples.

The 2 times table is easy because you simply double the results of the 1 times table.

With the children's help, fill in the second row and column of multiples.

To do the 4 times table, double the results in the 2 times table. And the 8 times table is double the results in the 4 times table.

Fill in the fourth and eighth rows and columns.

The 3 times table is simply the sum of the 2 times table and 1 times table. And the numbers in the 6 times table are double those in the 3 times table.

Fill in the third and sixth rows and columns.

You might remind children that multiplying 1 by a number gives the same answer as multiplying that number by 1 (see It all adds up on p44–5).

You can confirm the sequence of numbers in the × 3 and × 6 tables by counting in equal steps: 3, 6, 9, 12, 15... or 6, 12, 18...

To multiply 10 by any number, you just need to move that number one place to the left and put a zero in the ones place. And the 5 times table is always half the 10 times table.

With the children's help, complete the tenth and fifth rows and columns.

To find the numbers in the 9 times table, add the numbers in the 6 and 3 times tables.

Complete the ninth row of multiples.

This only leaves one missing number: the answer to 7 multiplied by 7, which is 49.
So you only have to learn one multiplication fact!

Each time you complete a table there are fewer and fewer gaps, as the products have been filled in when doing earlier tables.

pairs

Ask the children to discuss how they might use the completed table information to find an easy way to do the 11, 20, 19 or 16 times tables.

plenary

Add four extra rows to the bottom of the 10 × 10 grid OHT:

11									
16									
19									
20									

Gather in children's ways of doing the multiplications, and fill in the cells of the grid with their answers. For any multiplications that they can't do, offer suggestions:

- *To find a product in the 16 times table, double the equivalent product from the 8 times table. For example, 16 × 6 is 8 × 6, doubled.*
- *To find a product in the 19 times table, work out the equivalent product from the 20 times table and subtract the equivalent in the 1 times table. For example, 19 × 3 is 20 × 3 minus 1 × 3 or 60 minus 3, which is 57.*

Keep this part of the lesson fairly short. Move children onto discussing the other tables if you think they may have difficulties with one in particular.

You could prepare an extended version of the 10 × 10 multiplication grid that includes rows and columns for the 11, 16, 19 and 20 times tables – with the values up to 10 × 10 filled in. However, drawing only these rows is easier, and will suffice.

Children may have a range of strategies. Accept any that work, but focus mainly on those that make use of the rows of already completed facts, as in the main part of the lesson.

Something for the children to work on at home

Write out those single-digit multiplications that you still find hard to remember. Use the methods you have just learned to work out the answers.

Testing times

purpose	learning objective	equipment
✖ testing multiplication facts and establishing which ones children still need to learn	✖ know multiplication facts for all times tables to 10 × 10	✖ Resource sheet 1 for each child ✖ Resource sheet 6 or whiteboard grids for each child ✖ 1–9 number cards ✖ labelled cassette tapes, each with a test containing a mixture of questions from 1 × 1 up to 10 × 10, instruction poster, tape recorder and headphones

These two pages contain a range of tests to pick and choose from.

class test (all working on the same times table)

Present a 5 × 4 grid showing the products from a given multiplication table mixed up with other numbers. Arrange the multiples in some sort of pattern (such as alternate squares). Children copy this onto squared paper or whiteboard grids.

2	3	8	9	31
21	19	24	11	30
20	6	4	27	26
15	13	12	23	18

Call out multiplications from the times table you are testing; children find the product on their grid and shade it in (or put a counter on it). At the end, everybody should have the same pattern on their grid.

2	③	8	⑨	31
㉑	19	㉔	11	㉚
20	⑥	4	㉗	26
⑮	13	⑫	23	⑱

Ask children to note the multiplications they still need to work on.

As a follow-up, ask children to devise a grid with a different pattern for, say, the 5 or 9 times table. Use one of these next time you have a times table test.

class test (all working on the same times table)

Give children a copy of Resource sheet 1 and allow them three minutes to practise the times table in question. Now ask the children to turn their grids over. Then call out the questions, not in order, while the children list their answers. At the end, children swap sheets. Write up the answers on the board and ask children to mark each other's work.

Go over any of the multiplications the children have not yet memorised.

class test (different times tables at the same time)

Children need a 2 × 5 grid on squared paper or whiteboards with these operations written in the cells and space for their answers.

×1	×2	×3	×4	×5
×6	×7	×8	×9	×10

Give each group of children a number card showing the table you are testing them on; they keep this turned face down, so they don't know which number it is.

The first time, give children 30 seconds to fill in their grid, then insist 'pencils down'. Next time, try 20 seconds or less.

Now ask everybody to turn over the card, and begin filling in the answers to the multiplications on their grid.

After the allotted time, children swap sheets. Ask the children to use Resource sheet 1 to mark each other's work.

Any child who gets all the answers right can move on to a different table next week.

individual tests

Set up a quiet corner where children can sit with a tape recorder, listen to the tapes, and test themselves. Display instructions so that the children know how to test themselves:

> Testing your tables
>
> Each tape has ten questions on it from different times tables.
>
> Choose which tape you want to listen to and put it in the tape recorder.
>
> Prepare an answer sheet by numbering your paper 1 to 10.
>
> Put the headphones on, and switch on the tape recorder.
>
> On your sheet, make a note of the time you start and the time you finish.
>
> Hand in your completed sheet for marking.
>
> Rules
>
> No working out is allowed.
>
> Only write down the answers.
>
> Use the pause button if you need more time to answer a question.

Children can have as many attempts as they like, when they like, without disturbing anyone else in the class.

Include a mixture of 'easy' and 'hard' questions in each test, in order to create some sort of balance and equivalence between the tests. You can use a mixture of presentation styles ('six 7s', 'the product of 6 and 7', '6 multiplied by 7', 'seven lots of six'), or choose one style and stick to it.

Explain to the class that the tests contain a mixture of multiplication questions, as it is important that they can give answers to questions that are out of the context of a particular times table.

Children keep a record of the multiplications they still need to work on, perhaps by listening to a tape again and noting the questions they couldn't do.

> **Something for the children to work on at home**
> Practise the multiplication facts that you still find difficult to remember, using Resource sheet 1 as a reminder.

It all adds up

purpose
- using the multiplication grid to find products up to 19×19

learning objectives
- know multiplication facts for all times tables to 10×10
- partition

equipment
- OHT of Resource sheet 1 and copies for each child
- four strips of paper the length and width of a row or column of the multiplication grid
- calculators for each pair

This activity builds on children's experience of using the multiplication grid to find products up to 10×10. They learn to split the numbers in multiplications such as 15×13 into parts, thus modelling the process of partitioning.

Children may know how to use the grid effectively; but putting this knowledge into words can help them develop their understanding of the principles involved.

The square numbers are an exception as they appear only once. For example, 7×7 swapped around is still 7×7, so its product always appears in the same place.

class and pairs

Display the OHT of Resource sheet 1. Ask pairs to agree on an explanation of how the grid works that would make sense to someone who had never seen one.

Collect in ideas, and establish a common explanation. Draw out the generalisation that the product of two numbers is found at the intersection of the row and column headed by those numbers.

Establish that, except for the squares, each product appears twice. For example, the product of 4 and 8 appears eighth in the '4' column and fourth in the '8' row. Demonstrate this by covering up the numbers in the '4' row with one strip of paper and the numbers in the '8' column with another strip of paper. Point to the intersection and ask what number is underneath that square.

×	1	2	3	4	5	6	7	8	9	10
1	1	2	3	4	5	6	7		9	10
2	2	4	6	8	10	12	14		18	20
3	3	6	9	12	15	18	21		27	30
4										
5	5	10	15	20	25	30	35		45	50
6	6	12	18	24	30	36	42		54	60
7	7	14	21	28	35	42	49		63	70
8	8	16	24	32	40	48	56		72	80
9	9	18	27	36	45	54	63		81	90
10	10	20	30	40	50	60	70		90	100

Move the strips to the '8' row and the '4' column. Establish that the same number is underneath.

Explain that we can use this grid to find the products of numbers greater than 10. For this, we need to pay attention to the intersecting squares.

Show how it works with 13×16. Ask one child to multiply the two numbers on a calculator but to keep the answer secret. Put rings around the column headings 10 and 3 and the row headings 10 and 6.

What have 10 and 3 and 10 and 6 got to do with 13 multiplied by 16?

Use two strips of paper to cover the numbers in the '10' and the '3' rows. Use the remaining strips to cover the numbers in the '10' and the '6' column.

Ask what number lies underneath each of the four intersecting squares. Write these on the board and find their total. Compare this total with the multiplication done on the calculator.

Do you think this will work for other multiplications?

Repeat the process as many times as you think appropriate, using other multiplications.

18
30
60
100
208

×	1	2	3	4	5	6	7	8	9	10
1	1	2	3	4	5		7	8	9	
2	2	4	6	8	10		14	16	18	
3										
4	4	8	12	16	20		28	32	36	
5	5	10	15	20	25		35	40	45	
6	6	12	18	24	30		42	48	54	
7	7	14	21	28	35		49	56	63	
8	8	16	24	32	40		56	64	72	
9	9	18	27	36	45		63	72	81	
10										

pairs

Each child writes down five multiplications greater than 10 × 10 but less than 20 × 20 for their partner to do. When the partner has worked out the answers, the first child checks the multiplications using a calculator.

This is providing further practice of the multiplication facts up to 10 × 10.

plenary

Write down a multiplication in secret and say that you want the children to work out what it is and its answer. Use the four strips of paper to cover up two columns and two rows on the OHT grid but this time don't use the 10 row.

×	1	2	3	4	5	6	7	8	9	10
1	1	2	3	4	5	6			9	10
2	2	4	6	8	10	12			18	20
3	3	6	9	12	15	18			27	30
4										
5	5	10	15	20	25	30			45	50
6										
7	7	14	21	28	35	42			63	70
8	8	16	24	32	40	48			72	80
9	9	18	27	36	45	54			81	90
10	10	20	30	40	50	60			90	100

Suppose you write down 13 × 12. Instead of breaking this down into 10 + 3 and 10 + 2, split the numbers in a different way. Try, 7 + 6 and 8 + 4, and cover the relevant rows and columns.

These are all the clues I am going to give you.

What two numbers am I multiplying?

What is their product?

What other columns and rows could I have covered to find the answer to my multiplication?

You can cover the 9 and 4 rows and the 5 and 7 columns, in fact, any pair of rows and any pair of columns that add up to the numbers you are multiplying.

Something for the children to work on at home

- Practise using Resource sheet 1 to multiply other pairs of numbers below 20.
- Explore how the grid could be used to multiply numbers up to 30, 40 or 50.
- Investigate covering up three rows and three columns to multiply a pair of numbers such as 18 and 15 or 21 and 12.

Alphabet soup

purpose
- looking for patterns in multiplication tables

learning objectives
- explain methods and reasoning
- solve mathematical problems and puzzles
- recognise and explain patterns and relationships

equipment
- OHTs of Resource sheets 22, 23 and 24
- Resource sheet 24 for each pair

Noticing the patterns of numbers in the times tables is a great help when learning them; and pattern awareness is an important mathematical ability to develop. This lesson helps children focus on the patterns of the times tables through codes, by turning the digits into letters.

class

Display the OHT of Resource sheet 22. Discuss patterns in the answers to the 5 times table.

> The ones digits alternate, and there are only two different digits (0 and 5). In the tens column there are two 1s, two 2s, two 3s, two 4s and a 5.
>
> The letter 'I' could be confused with the digit '1'.

individuals

Children now write down the digits 0 to 9 in a line. Underneath, they write down the letters A to K, omitting the letter I, in any order, thus producing a code.

0	1	2	3	4	5	6	7	8	9
H	A	K	D	C	B	F	J	G	E

Now the children write out the 5 times table up to 5 × 9, then copy it out again but this time replace the numbers with the corresponding letters from their code.

class

Display the coded E times table from Resource sheet 23 (the 5 times table up to 5 × 9) and ask the class which table it is. They may quickly recognise it as another version of the 5 times table.

> E × J = E
> E × B = JC
> E × F = JE
> E × G = BC
> E × E = BE
> E × H = FC
> E × A = FE
> E × D = GC
> E × K = GE

Does it look like any of your coded times tables?

How can you tell that this is the 5 times table?

> Because the table is in order, children can work out that the fifth equation is E × E, so E is 5, and the code is not hard to crack:
>
> 1 2 3 4 5 6 7 8 9 0
>
> J B F G E H A D K C

Give them a couple of minutes to start to decode it (but don't expect a thorough check at this stage). Go over the first few coded equations with them.

E must be 5. So in the first equation, what is J?

Look at the second equation. What must B and C be?

Next, show the scrambled version of the A times table from Resource sheet 24. Explain that it is also the 5 times table, but in a different code, and with the equations not in the usual order.

How can you still tell it is the 5 times table?

> A × K = JC
> A × J = EA
> A × G = JA
> A × E = A
> A × F = HC
> A × B = EC
> A × D = HA
> A × H = BC
> A × A = BA

> In the scrambled version:
> - the answers all end in either C or A
> - the square of 5 (A) ends in a 5 (A)
> - in the tens column of the answers, there are two entries for each letter
> - there is only one multiple of 5 that is a single-digit answer

pairs

Show the scrambled version of the B times table (the 1 times table). Explain that it is a different table, with a different code, and with the equations not in the usual order.

Give pairs two minutes to discuss what table it is.

class

Bring the class together.

Which times table do you think it is? Why do you say that?

Now show the F times table from Resource sheet 23 (the 9 times table in order, again with a new code.)

This times table is in order. Which do you think it is?

As it is in order, we can work out what B must be. It must be 1. Where else does B appear?

Sort out a few of the equations with the class, but don't crack the code completely.

$B \times A = A$
$B \times F = F$
$B \times G = G$
$B \times C = C$
$B \times B = B$
$B \times K = K$
$B \times D = D$
$B \times J = J$
$B \times H = H$

$F \times B = F$
$F \times H = BJ$
$F \times A = HG$
$F \times K = AC$
$F \times E = KE$
$F \times C = EK$
$F \times G = CA$
$F \times J = GH$
$F \times F = JB$

There is only one number which, when multiplied by another number, gives that new number. This must be the 1 times table.

$1 \times$ any number $=$ that number

However as there are no further clues it is not possible to crack the code.

F is the ninth multiple so it must be 9:

$F \times B = F$

$9 \times 1 = 9$

$F \times H = BJ$

$9 \times 2 = 18$

pairs

Hand out Resource sheet 24. Children try to identify the C and D tables and agree reasons for their thinking.

Early finishers can start to crack the codes of all the times tables on Resource sheet 24, writing out the correct equations underneath the coded ones.

The C and D times tables on Resource sheet 24 are the 3 and 4 times tables, respectively.

plenary

Bring the class together and ask for solutions, strategies and reasons for the choices made.

Look at the OHT of the 3 and 4 times tables and compare these with the coded versions to confirm the identity of the coded tables.

In the 3 times table, how many answers only have one digit?

Is that true of any other times table?... So is this coded table the 3 times table?

In the coded times table, which line must be 3 × 3? So what is F worth?

$3 \times 1 = 3$	$C \times K = GK$	$4 \times 1 = 4$	$D \times C = D$
$3 \times 2 = 6$	$C \times F = BJ$	$4 \times 2 = 8$	$D \times J = GE$
$3 \times 3 = 9$	$C \times B = D$	$4 \times 3 = 12$	$D \times H = GB$
$3 \times 4 = 12$	$C \times J = BG$	$4 \times 4 = 16$	$D \times F = CG$
$3 \times 5 = 15$	$C \times C = F$	$4 \times 5 = 20$	$D \times E = FG$
$3 \times 6 = 18$	$C \times G = C$	$4 \times 6 = 24$	$D \times G = E$
$3 \times 7 = 21$	$C \times D = GA$	$4 \times 7 = 28$	$D \times D = CA$
$3 \times 8 = 24$	$C \times H = GB$	$4 \times 8 = 32$	$D \times A = GD$
$3 \times 9 = 30$	$C \times A = BH$	$4 \times 9 = 36$	$D \times K = FA$

Solutions

A times table:
1 2 3 4 5 6 7 8 9 0
E B J H A K G F D C

C times table:
1 2 3 4 5 6 7 8 9 0
G B C H K D J A F E

D times table:
1 2 3 4 5 6 7 8 9 0
C G F D H A J E K B

Something for the children to work on at home

Continue cracking the codes in the scrambled tables.

Put into code, then scramble, the 9 or 6 times table for a friend to identify and decode.

Factoring in pairs

purpose	learning objectives	equipment
exploring factors and factor pairs	find all the pairs of factors of any number up to 100 factorise numbers to 100 into prime factors recognise and explain patterns and relationships	counters squared paper an empty factor grid on an OHT or to put on the board sticky tape

It is important that children see the patterns underlying the number system. The factors table introduced in this activity demonstrates the regularity of factors, and the relationships with their products.

class

Put ten counters on the OHP. Discuss all the ways that the ten counters can be arranged in equal groups, and record these on the board under the heading '10'. Make sure to include 10 and 1.

10
1
2
5
10

The counters can be arranged as a group of ten, and also as ten ones.

Introduce the word 'factor' if this is unfamiliar to the children.

If you can arrange a number in equal groups, the number in that group is a factor. So 10 has these factors: 10, 5, 2 and 1.

Can anyone tell me the factors of 6... and of 5?

Ask children to think for a moment about how 9 can be arranged in equal groups.

Any idea how nine counters can be arranged in equal groups? In other words, what are the factors of 9?

Test out the children's ideas on the OHP with nine counters.

> **You can use an OHT of Resource sheet 6 to help you draw this table.**

Display this table on the board or OHP, and put ticks in the cells to show the factors of 10 and 9.

Explain that it can provide a good way of listing the factors of all the numbers from 1 to 10. Ask which numbers are factors of 8, and fill these in.

Numbers

	1	2	3	4	5	6	7	8	9	10
1									✓	✓
2										✓
3									✓	
4										
5										✓
6										
7										
8										
9									✓	
10										✓

individuals

Children stick together sheets of squared paper to make a grid containing at least 25 × 25 squares. In the top left-hand corner of their paper they compile a table similar to the one on the board/OHP and fill it in. For homework they will extend the table downwards and to the right.

> **You can provide children with ready-made tables to complete to save the time it would take to reproduce the table from the board.**

plenary

Ask the children to use their own work to help you complete your table. Quick-fire some questions about the data it contains.

What are the factors of 8?

What do we call numbers that have 2 as a factor?

What do we call numbers that have only got two factors?

Together, add on rows and columns for 11 and 12, and fill them in:

Numbers

	1	2	3	4	5	6	7	8	9	10	11	12
1	✓	✓	✓	✓	✓	✓	✓	✓	✓	✓	✓	✓
2		✓		✓		✓		✓		✓		✓
3			✓			✓			✓			✓
4				✓				✓				✓
5					✓					✓		
6						✓						✓
7							✓					
8								✓				
9									✓			
10										✓		
11											✓	
12												✓

Ask the class to focus on column 12. Invite individuals to use the data in the table to find two numbers that multiply together to give 12. Write these on the board. Keep going until you have all the equations on the board.

$4 \times 3 = 12$
$3 \times 4 = 12$
$1 \times 12 = 12$
$12 \times 1 = 12$
$2 \times 6 = 12$
$6 \times 2 = 12$

Ask the children to imagine the grid on the OHP/board being extended along and upwards.

Imagine the column with 20 at the top. Which numbers will be ticked in that column? Why do you think that?

How could you make sure you had them all?

In answer to the last question, demonstrate the strategy of 'pairing' factors starting with 1 and building upwards until you 'meet in the middle'.

20
1   2   4   5   10   20
1 and 20, 2 and 10, 4 and 5

Ask the children to use these factor pairs to produce some multiplication equations with an answer of 20.

Finally, ask them to find the factors of 24 and of 25, and try the pairing technique. Discuss how, with square numbers, such as 25, there is always one factor pair that contains just one factor, for example, 5×5, and so there is an uneven number of factors.

Provide counters or cubes for modelling. Ask the children to fill in one column at a time and then check with you.

The factors of 8 are 1, 2, 4 and 8.

The numbers in this row are even.

Numbers with only two factors are called prime numbers. These numbers will only have two ticks against them.

You may wish to discuss whether, in this context, to count pairs such as 12×1 and 1×12 as the same as or different to each other.

The factors of 20 are: 1, 2, 4, 5, 10 and 20.

Mapping factors onto a table, as in this activity, is one way to establish factors. Another way, not addressed here, is to divide the number in question by all potential factors from 1 upwards, in turn, until you reach halfway. If the result is a whole number, that number is a factor.

The factors of 25 are 1, 5 and 25

The factors of 24 are: 1, 2, 3, 4, 6, 8, 12 and 24.

> ### Something for the children to work on at home
> Continue your tables as far as 25, then answer these questions:
> Which number less than 26 has the most factors? (The answer is 24.)
> Which number less than 50 do you think has the most factors? How could you check? (The answer is 48.)

Equal shares

purpose	learning objective	equipment
exploring and modelling division	extend understanding of the operations of multiplication and division, and their relationship to each other	OHP calculator individual whiteboards and pens counters for the OHP

Children may be familar with an array of squares, circles or counters as a way of representing multiplication. Here the array is used to explore division and to generate division equations.

class

Display the following array on the board, and establish that it shows twelve circles (also three rows of 4 and four columns of 3).

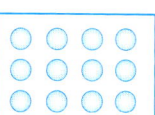

Ask the following questions, making sure that the children listen carefully to the language. Volunteers demonstrate the correct interpretation each time by drawing on the diagram. Ask for suggestions about how to do each relevant calculation on the OHP calculator.

You will need to rub out the circles children draw on the board after each turn or present four versions of the array on the board.

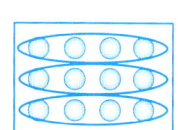

Circle each row of 4.

On the OHP calculator do

$12 \div 3 =$

When 12 is divided into three equal groups, how many are there in each group?

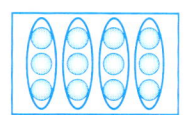

Circle all 4 columns of 3.

On the OHP calculator do

$12 \div 3 =$

When 12 is divided into equal groups of 3, how many groups are there?

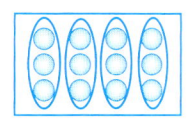

Circle each column of 3.

On the OHP calculator do

$12 \div 4 =$

When 12 is divided into four equal groups, how many are there in each group?

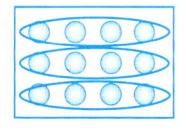

Circle each row.

On the OHP calculator do

$12 \div 4 =$

When 12 is divided into equal groups of 4, how many groups are there?

Ask children to draw an array on their whiteboards that shows 12 divided into equal groups of 2.

$12 \div 2 =$

What division shall I do on my calculator to check how many equal groups of 2 there are?

Now ask children to draw an array on their whiteboards that shows 12 divided into two equal groups.

$12 \div 2 =$

What division shall I do on my calculator to check how many there will be in each equal group?

Discuss the question:

What does 12 ÷ 6 mean?

pairs

Ask the children to draw arrays to illustrate the following divisions:

20 ÷ 5   18 ÷ 6   14 ÷ 7   24 ÷ 8   27 ÷ 9

plenary

Bring the class together and discuss their diagrams and divisions.

Can you read out this equation?

Can you describe to us a diagram you drew?

Does that sound like the diagram that other people drew?

Now write 91 ÷ 7 = 13 on the board. Explain that you are going to ask a number of questions. Some of these can be answered by using this multiplication equation and others can't. If the children can answer the question using the information in the equation, they write the answer on their whiteboards, and hold them up to face you. If they can't answer the question, then they just write a big cross (meaning 'can't be answered').

| 91 ÷ 7 = 13 |

How many equal groups of 7 can I make out of 91? (13)

When I divide 91 into seven equal groups, how many are there in each group? (13)

What is 91 multiplied by 13? (can't be answered)

What is 91 divided by 13? (7)

When I divide 91 into thirteen equal groups, how many are there in each group? (7)

When I divide 13 into groups of 7, how many groups do I have? (can't be answered)

What is 13 multiplied by 7? (91)

What is 7 multiplied by 13? (91)

What is 91 multiplied by 7? (can't be answered)

What is 7 divided by 13? (can't be answered)

> **Something for the children to work on at home**
> Draw a 9 × 6 array of squares, then write two addition equations, two multiplication equations and two division equations that relate to it.

Make sure that children understand that there are two interpretations: 12 divided into equal groups of 6, and 12 divided into six equal groups.

Aim for two arrays for each division but accept just one.

9 + 9 + 9 + 9 + 9 + 9 = 54

6 + 6 + 6 + 6 + 6 + 6 + 6 + 6 + 6 = 54

9 × 6 = 54

6 × 9 = 54

54 ÷ 9 = 6

54 ÷ 6 = 9

Long activities

The missing link

purpose	learning objective	equipment
recognising and using the links between multiplication and division	use the relationship between multiplication and division	OHP calculator calculators for each pair piece of card with a question mark on the front

This activity aims to reinforce in children's minds the links between multiplication and division, and to give them practice in turning one type of statement into another.

class

On the board, write 7, 3 and 21.

| 7 | 3 | 21 |

These three numbers are connected in some way. What do you think the connection is?

Accept all valid ideas but home in on $7 \times 3 = 21$

Write the same three numbers again, in a different order, leaving a space between the numbers.

I wonder if anyone can think of a way to write those three numbers in a different order?

Keep going until you have all six combinations.

7	3	21
7	21	3
3	7	21
3	21	7
21	3	7
21	7	3

Explain that you can insert symbols in the spaces between the numbers to make true mathematical statements (or equations). The three symbols to choose from are $\times$, $\div$ and $=$.

Enter the signs for the first equation, and then complete one of the more 'difficult' ones.

Children may never before have seen an equation with the equals sign before the operation sign.

Work together as a class to complete the rest. Two of the examples have two solutions (one using the multiplication and one using the division symbol). So when you have a complete list of six equations, there will be two equations which could be written out differently. For each one, ask what the other solution is.

These are

21 3 7 and 21 7 3

$21 = 3 \times 7$ $21 = 7 \times 3$

$21 \div 3 = 7$ $21 \div 7 = 3$

This one [21 3 7] has a multiplication sign. How could we use a division sign instead?

For 21, 7 and 3 we have $21 = 7 \times 3$. How else could we fill in the gaps?

| $7 \times 3 = 21$ |
| $7 = 21 \div 3$ |
| $3 \times 7 = 21$ |
| $3 = 21 \div 7$ |
| $21 = 3 \times 7$ |
| $21 \div 3 = 7$ |
| $21 \div 7 = 3$ |
| $21 = 7 \times 3$ |

pairs

Ask a child to give you two different digits and their product. Write these on the board.

Remind the children that there will be two solutions for some equations.

(There will always be two solutions where the product comes first.)

Pairs write out all six combinations of these three numbers and then insert the $\times$, $\div$ or $=$ symbols to make true equations.

Times Table Tactics

class

Ask the children to help produce a list of solutions on the board, including two for each of the combinations where the largest number came first.

Cover up any one of the numbers by sticking a piece of card marked with a question mark on the front over it.

If I didn't know the number under the card, how could I use the calculator to obtain the missing number?

Can I press 30 ÷ ? = 6 on the calculator?

$5 \times 6 = 30$
$5 = 30 \div 6$
$30 = 5 \times 6$
$30 = 6 \times 5$
$30 \div ? = 6$
$30 \div 6 = 5$

The card marked with a question mark allows you to replicate missing number problems.

Discuss how to use the two visible numbers to work out the answer, and use the OHP calculator to check the ideas suggested.

In turn, cover up the different numbers in the same equation. Then cover up numbers in the other equations.

Sometimes the equation can simply be keyed into the calculator:
$5 \times 6 = ?$
More often, the equation will need to be rearranged:
$? = 5 \times 6 \rightarrow 5 \times 6 =$
And the operation sign may need to change:
$30 = ? \times 6 \rightarrow 30 \div 6 =$

pairs

Each pair needs a calculator. Present these equations:

$? \div ? = ?$
$? = ? \times ?$
$? \times ? = ?$
$? = ? \div ?$

Pairs insert a different set of numbers in each equation so that it is true. When they have finished, they write out the four equations again but with one of the numbers missing from each one, and swap these 'missing number' problems with another pair to solve.

More adventurous children can work with 'hard' calculations such as $342 \div 19 = 18$

But discourage the use of decimals:
$349 \div 19 = 18.368421$

This tests whether children can apply the process in unfamiliar situations.

plenary

Secretly, use a calculator to carry out some multiplications and divisions. Each time write up the equation, but leave out one of the numbers. As you do this, ask children to estimate the size of the missing number.

$57 \times ? = 2451$
$? \div 23 = 161$
$2296 \div ? = 82$
$? \times 18 = 342$

With the children, find the missing numbers using the OHP calculator.

57 multiplied by something equals 2451. Roughly, how big is that something? How can I use the calculator to find the missing number?

Something for the children to work on at home

Write out all the equations that connect these three numbers: 2, 7 and 14, or these three: 9, 6 and 54. Or pick your own trio of numbers.

Long activities

All or nothing

purpose	learning objectives	equipment
relating division to factors and factor pairs	find all the pairs of factors of any number up to 100 know division facts corresponding to tables up to 10 × 10 use the relationship between multiplication and division	OHT of Resource sheet 1 OHT or copies of Resource sheet 25 squared paper calculators for each pair OHT of a blank 13 × 13 grid from Resource sheet 6

Children readily use and become familiar with multiplication grids. This activity introduces the division grid, and makes links between it and the multiplication grid.

class

Display the OHT of Resource sheet 1.

On the board write:

$36 \div 9 = 4$

Look at the multiplication grid. How do the three numbers in this equation link together on the grid?

36 is in the intersection of the ninth column and fourth row and of the fourth column and ninth row.

Now write:

$42 \div 7 = ?$

How could you use the multiplication grid to find the answer to this division?

Look for 42. It is in the seventh row (or column) and it is the sixth number in the column (or row).
So $6 \times 7 = 42$ and $42 \div 7 = 6$

Write up another four or five similar divisions with the answer missing each time.

How could you use the multiplication grid to find these answers?

On the board write:

$? \div 5 = 9$
$40 \div ? = 5$

How could you use the multiplication grid to find the answer to these divisions?

Go to the fifth column, then down to the ninth row or go to the fifth row and along to the ninth column, and read the number. Use the same method, referring to the relevant columns and rows, for the second division.

Write up another four or five similar divisions with the answers missing.

How could you use the multiplication grid to find these answers?

Ask four or five children to suggest similar questions with the divisor missing, and solve these as a class.

class

Present a blank 13 × 13 grid on the OHP or board, and put the division sign in the top left-hand corner. Fill in the rest of the first row and the rest of the first column with the numbers 1 to 12.

Explain that this is a division grid, in some ways similar to the multiplication grid. To find the value for each cell you look for the number at the head of the column, and divide it by the number of that row.

What number belongs here? [the cell 2 ÷ 2] And here? [the cell 2 ÷ 1]

Ask which cells should have a 1 in.

Any number divided by itself gives 1, so 1 belongs in all of those cells.

Look at some cells where the answer would not be a whole number, such as 3 ÷ 2 or 5 ÷ 8. Ask children to do a few of these on a calculator, and establish that if a division does not produce a whole number answer, then the answer is a decimal number. Agree to put a cross in cells where the answers would be a decimal.

pairs

Children draw their own grids and fill them in up to 12 ÷ 12, using calculators where necessary.

Children who finish early can extend their grids to 15 ÷ 15.

Use grids large enough to go up to 20 ÷ 20, so that children can extend their grids later, as suggested for homework.

plenary

Present Resource sheet 25 and discuss how to use it to find various division facts.

What is 9 divided by 3?

What are the factors of 8?

And of 10?

Tell me some divisions with 2 as the answer.

Children can use the information from a multiplication grid for those number facts they do not yet know.

Find the column headed with the right product, and go down the column until you find a number written in it. The number is a factor of that product, and its pair is the number of that row.

÷	1	2	3	4	5	6	7	8	9	10	11	12	13	14	15
1	1	2	3	4	5	6	7	8	9	10	11	12	13	14	15
2	×	1	×	2	×	3	×	4	×	5	×	6	×	7	×
3	×	×	1	×	×	2	×	×	3	×	×	4	×	×	5
4	×	×	×	1	×	×	×	2	×	×	×	3	×	×	×
5	×	×	×	×	1	×	×	×	×	2	×	×	×	×	3
6	×	×	×	×	×	1	×	×	×	×	×	2	×	×	×
7	×	×	×	×	×	×	1	×	×	×	×	×	×	2	×
8	×	×	×	×	×	×	×	1	×	×	×	×	×	×	×
9	×	×	×	×	×	×	×	×	1	×	×	×	×	×	×
10	×	×	×	×	×	×	×	×	×	1	×	×	×	×	×
11	×	×	×	×	×	×	×	×	×	×	1	×	×	×	×
12	×	×	×	×	×	×	×	×	×	×	×	1	×	×	×
13	×	×	×	×	×	×	×	×	×	×	×	×	1	×	×
14	×	×	×	×	×	×	×	×	×	×	×	×	×	1	×
15	×	×	×	×	×	×	×	×	×	×	×	×	×	×	1

Establish how to use the grid to find pairs of factors for a given product.

Finish with a game. Children need their division grid to refer to (for numbers up to 12 or 15) and a calculator (for higher numbers). They work in twos or threes.

Call out a product from 1 to 50. Children try to find a pair of factors and write these on their whiteboards. Meanwhile, you choose a pair yourself and write these on your own whiteboard. When you say so, everybody holds up their whiteboard. Children with the same factors as you score 1 point; those with a different pair of factors score 2 points. Continue until one pair has scored 10 points.

With numbers that are not on their grids, children can always find at least one pair: the number itself and 1. And with even numbers, they can divide by 2. They can test other hunches on their calculators.

> **Something for the children to work on at home**
> Continue your grid as far as 15 ÷ 15 or 20 ÷ 20.

Number chains

purpose	learning objectives	equipment
practising multiplying by 2, 3, 4, 5, 7 and 8	know multiplication facts for the 2, 3, 4, 5, 7 and 8 times tables recognise and explain patterns and relationships generalise and predict	calculators for each pair (optional)

Investigating the behaviour of numbers is worthwhile in its own right, but here the investigation is also a great way to practise multiplication.

class

On the board write 2, 3, 4, 8, 7, 5 and tell the children you are going to do an investigation on each of those numbers in turn. Explain that there is a reason why the numbers are in that order.

We are going to start investigating the 2. The procedure is:

- *multiply the units by 2*
- *add on the number of tens (if any)*

We will keep applying this procedure to each successive answer.

Write down 2 and carry out the procedure until you reach 16, recording as you go.

Two multiplied by 2 is 4; add on no tens; we still have 4.

Four multiplied by 2 is 8; add on no tens again; we still have 8.

Eight multiplied by 2 is 16; add on no tens to get 16.

Six (the number of units) multiplied by 2 is 12, add on 1 (the number of tens) to get 13.

$$2 \times 2 + 0 = 4$$
$$4 \times 2 + 0 = 8$$
$$8 \times 2 + 0 = 16$$
$$6 \times 2 + 1 = 13$$

$2 \rightarrow 4 \rightarrow 8 \rightarrow 16 \rightarrow 13 \rightarrow$

You may want to carry on past 2, hoping the children will spot the repetition.

With the help of the children, continue the process until you get back to 2.

$2 \rightarrow 4 \rightarrow 8 \rightarrow 16 \rightarrow 13 \rightarrow 7 \rightarrow 14 \rightarrow 9 \rightarrow 18 \rightarrow 17 \rightarrow 15 \rightarrow 11 \rightarrow 3 \rightarrow 6 \rightarrow 12 \rightarrow 5 \rightarrow 10 \rightarrow 1 \rightarrow 2$

Accept all observations but home in on the facts:

- **all the multiples of 2 from 2 to 18 appear somewhere in the cycle**
- **the cycle contains all the counting numbers from 1 to 18 once and once only**

Have you noticed anything about the numbers in the cycle?

Do you think the same will happen with 3, the next number on our list? We will multiply the units by 3 and add on the number of tens, then keep going like that. Will we get back to 3?

pairs

Start the children off on the process by producing the first few terms together:

$3 \times 3 + 0 = 9$
$9 \times 3 + 0 = 27$
$7 \times 3 + 2 = 23$

Now ask them to complete the cycle.

class

Ask children to help you write out the complete cycle on the board.

$3 \rightarrow 9 \rightarrow 27 \rightarrow 23 \rightarrow 11 \rightarrow 4 \rightarrow 12 \rightarrow 7 \rightarrow 21 \rightarrow 5 \rightarrow 15 \rightarrow 16 \rightarrow 19 \rightarrow 28 \rightarrow 26 \rightarrow 20 \rightarrow 2 \rightarrow 6 \rightarrow 18 \rightarrow 25 \rightarrow 17 \rightarrow 22 \rightarrow 8 \rightarrow 24 \rightarrow 14 \rightarrow 13 \rightarrow 10 \rightarrow 1 \rightarrow 3$

What do you notice about the numbers in this cycle? Did you get back to 3?

How is this cycle like the previous one – and how is it different?

The cycle returns to 3, but contains more numbers than the 2 cycle.

- all the multiples of 3 from 3 to 27 appear somewhere in the cycle
- all the counting numbers from 1 to 28 appear in the cycle

pairs

Investigate what happens when they apply a similar process to 4, 8 and 7. They should tackle the numbers in that order.

plenary

Share the children's findings.

Do all the cycles return to their starting number?

How many multiples of 4 does the 4 cycle contain? What about the 8 cycle? And the 7 cycle?

What else have you noticed about the numbers in the 4 (or 8 or 7) cycles?

The 4 cycle returns to 4 after six operations and contains only two multiples of 4: itself and 16.

The 8 cycle returns to 8 after 13 operations and contains only two multiples of 8: itself and 64.

The 7 cycle returns to 7 after 22 operations and the list contains only three multiples of 7: itself, 28 and 49. It also contains a sequence of numbers that go up by 3 each time: 1, 4, 7, 10, 13... to 67, but does not include 46.

Something for the children to work on at home

Complete the cycle starting with 5. Then answer these questions:

How many numbers are there in the cycle?

Does the cycle go back to 5?

The cycle contains all the whole numbers from 1 to 48 except for a particular group. Which group is that?

The cycle contains all the multiples of 5 from 5 to 45, except one. Which one?

There are 42.

Yes.

Multiples of 7 from 7 to 42 are the missing group.

35 is the missing multiple of 5.

Bite size pieces

purpose
practising multiplying by 6, 8 and 9

learning objectives
know multiplication facts for the 6, 8 and 9 times tables

recognise and explain patterns and relationships

generalise and predict

equipment
OHT of Resource sheet 6

calculators (optional)

This activity helps children see connections between different times tables.

class

Present this table on the board or OHP. In the first column write the numbers 1 to 10 in a random order.

Number	× 3 and × 2	× 6
5		
1		
9		
3		
7		
6		
10		
2		
8		
4		

With the help of the children, take each number in turn, multiply it by 3, then double the result. Write the answers in the middle column.

Number	× 3 and × 2	× 6
5	15 30	
1	3 6	
9	27 54	
3	9 18	
7	21 42	
6	18 36	
10	30 60	
2	6 12	
8	24 48	
4	12 24	

Some children might need to use a calculator to help them with the calculations in the last column.

Cover up the middle column and ask individual children to multiply a number in the first column by 6. Enter the results in the last column (including any incorrect answers they may give). When the last column is complete, uncover the middle column.

Number	× 3 and × 2	× 6
5	15 30	30
1	3 6	6
9	27 54	54
3	9 18	18
7	21 42	42
6	18 36	36
10	30 60	60
2	6 12	12
8	24 48	48
4	12 24	24

Multiplying a number by 3 and then doubling the result is equivalent to multiplying it by 6.

What do you notice? Why do the answers in the last column also appear in the middle column?

Do you think the same thing would happen if there were numbers greater than 10 in the first column?

pairs

Pairs draw three columns with the same headings, write any five numbers greater than 10 in the left-hand column and, using a calculator if necessary, complete the other columns.

class

Bring the class together and discuss their findings.

Did the same thing happen? Did multiplying those bigger numbers by 3 and then by 2 have the same effect as multiplying by 6?

Tell the children that someone said a quick way of multiplying by 9 is to multiply by 3 and then multiply the result by 3.

Do you think that is true?

How could you check it?

If we used a table similar to the one we had earlier, what would the headings be?

Once you are satisfied that they know what the task is, let them investigate first using the numbers 1 to 10 (in random order) and then using five numbers greater than 10.

plenary

Present the OHT of Resource sheet 5. This is to help develop a physical model of multiplying by 3 then doubling, and of multiplying by 6. Invite three children to the front of the class. One child shades a row of, say, five squares near the top of the grid, then announces and records the number.

I've shaded five squares.

The next child trebles the number of shaded squares, by shading another two rows. They also announce the number of shaded squares and record it.

There are fifteen shaded squares.

The third child doubles the number of shaded squares. They announce the final number of shaded squares and record it.

There are 30 shaded squares.

Now shade a row of five squares, then 'multiply' this by 6.

Establish that the two lots of shaded squares are equal in number.

That row of five squares was multiplied by 3 and then by 2, and it made 30 squares. This row of five squares was multiplied by 6, and it also made 30 squares.

Repeat for multiplying by 3 then by 3 again, and then by 9.

> ### Something for the children to work on at home
> Write down any number. Multiply it by 6 and then multiply it by 9, and look for a 'special relationship' between the two answers. Check whether or not you are right by testing some different starting numbers.

The method works for any number.

The headings would be:
Number   × 3 and × 3   × 9

Children can use a calculator to help.

You could have a fourth child (or the whole class) carry out the parallel calculations on a calculator.

5 × 3 × 2 = 30

5 × 6 = 30

Use another OHT sheet for this second example.

The larger product is half as much again as the smaller one.

Telephone numbers

purpose	learning objectives	equipment
practising multiplication through games	recognise multiples of 2, 3, 4, 5, 6, 7, 8 and 9 know multiplication facts for the 2–9 times tables	one set of cards cut from Resource sheet 26 for each group three sets of 0–9 number cards for each pair; five sets for each group 4 × 5 grids (from squared paper) for each child

Children need to practise using and recalling facts from the multiplication tables until they have them at their fingertips. Games and enjoyable activities such as these are an excellent way to do this.

class

Establish that, for these games, a 'telephone number' is four digits that come from the same complete multiplication equation. So 3927 (pronounced 'three nine two seven') is just one example and is derived from the equation 3 × 9 = 27.

Allocate these activities to pairs or groups, as appropriate:

Badges – for any number of children

The children make and wear badges with different 'telephone numbers' on them. During the day, children must refer to others in the class, not by name, but by the telephone number that that person is displaying. They can change their badges each day.

Variation: the badge shows only a two-digit number, say 39 (the first two digits). But that person is referred to by the completed four-digit telephone number: 3927.

Variation: the badge shows just a product, say 24. The wearer is referred to by any valid four-digit telephone number that 'works' (such as 4624 or 8324).

Telephone scores – for pairs

Each player has a 4 × 5 grid that fits five rows of four number cards. Each pair has a pack of number cards (three sets of 0–9). They shuffle these and place them face down in the centre of the table.

Player 1 picks a card and places it anywhere on their grid. Player 2 does the same on their grid. Play continues alternately. The objective is to make as many 'telephone numbers' as possible. Once a card is placed, it must remain in that position.

Each player is allowed up to four discards, where they set a card aside rather than place it on the grid.

The scoring system is based on the value of the third digit so, for example, 5735 scores three points and 9981 scores eight points. But 4208 is also a valid 'telephone number', and so scores one point too.

Some multiplications in the tables, such as 3 × 2 = 6 or 4 × 10 = 40, would make three- or five-digit telephone numbers: 326 or 41040

Tell children to stick to four-digit telephone numbers, as these are what the other activities require.

4624 and 8324 both 'work' because 4 × 6 = 24 and 8 × 3 = 24.

In 'Telephone scores' an equation such as 3 × 2 = 6 should be seen as 3 × 2 = 06, so the 'telephone number' is 3206 not 326.

for example:

3	9	2	7		3 × 9 = 27	score 2
3	6	1	5		3 × 6 ≠ 15	score 0
5	4	4	0		5 × 4 ≠ 40	score 0
4	1	0	4		4 × 1 = 04	score 1
2	6	1	2		2 × 6 = 12	score 1

discarded: 1 7 9 6

Variation: at the end, players rearrange their cards to make the best scores that they can.

Telephone rummy – for 4 children in pairs

The group has a pack of shuffled number cards (five sets of 0–9). They deal each pair five of these cards and place the rest face down to form the pick-up pile. They then take the top card from this pile and put it face upwards, to form the start of the discard pile.

Pair 1 complete the following three steps:

- *Pick a card from either the discard pile or the pick-up pile.*
- *Look to see if they can make a four-digit telephone number. If they can, they lay down those four cards in front of them. Now they pick up another four cards, to replace them, from the pick-up pile (only one set of four may be put down in each turn).*
- *Discard one card.*

Pairs should always have five cards at the end of their turn.

Pair 2 does the same. Play continues until the pick-up pile runs out. The winning pair is the one who lays down the most telephone numbers.

Players should put down four cards at a time. If they want to make a three-digit telephone number (say, 248) they should use zero in the tens place, and so need four cards:

2408 not 248

Equations involving two-digit numbers, such as $3 \times 10 = 30$ (31030) aren't possible in this game.

Pelmanism – for 2 to 4 children

The group needs a set of Question and Answer cards cut out from Resource sheet 26. They spread out the Question and Answer cards separately, face down. Each player, in turn, turns over a card from each group. If the cards match, the player takes them and has another go.

| 38 | 24 | 49 | 32 |

These match because $3 \times 8 = 24$.   These don't match: 4×9 does not equal 32.
If they don't match, the cards are turned over and play continues.

The winner is the player with the most telephone numbers.

The Pelmanism set covers two tables – the 8 and the 9 times tables. You, or the children, could make up sets for the 2, 3, 4, 5, 6 and 7 times tables.

plenary

Ask children for examples of the following four-digit 'telephone' numbers:

- *where all the digits are even (eg 4624)*
- *where all the digits are odd (eg 3515)*
- *whose digits are in descending order from left to right (eg 9872)*
- *whose third digit is 6 (eg 7963)*
- *whose last digit is 6 (eg 6636)*

Some children may need to refer to a multiplication grid.

Something for the children to work on at home

Make up cards for Pelmanism using any pair of times tables.

Play Pelmanism alone or Telephone scores with a partner.

Long activities **61**

ISBN numbers

purpose	learning objectives	equipment
introducing a real-life application of multiplication	know multiplication facts for the 2 to 9 times tables develop calculator skills and use a calculator effectively solve mathematical problems and puzzles	OHP calculator plenty of published books (subjects not relevant) Resource sheet 27 (as OHT or individual copies)

International Standard Book Numbers (ISBNs) use multiplication and division to give each book a unique reference number. This activity explores ISBNs to practise multiplication and division skills.

class

Ask if anyone knows what an ISBN number is.

Ask someone to read the ISBN number from any book (they should choose one without X as the check digit) but not to give you the last digit. Explain that when a shop receives an order for a book, a computer multiplies the first digit by 10, the second digit by 9, the third digit by 8 and so on. These products are then added together. The purpose of the last digit, the check digit, is to adjust the total so that it divides by 11 exactly.

An ISBN is an International Standard Book Number which is unique to, and identifies, each published book. The first nine digits indicate the language, the publisher and the book. The tenth digit is a check digit. Sometimes this check digit is an X, which stands for 10.

For example, 1 874099 33 (2)

$(1 \times 10) + (8 \times 9) + (7 \times 8)$
$+ (4 \times 7) + (0 \times 6) + (9 \times 5)$
$+ (9 \times 4) + (3 \times 3) + (3 \times 2)$
$= 262$

Together, apply the rule to the first nine digits of the ISBN number and work out the total.

We need to check whether this will divide by 11.

First, use the OHP calculator to generate multiples of 11 up to 220 or beyond; write these on the board.

| 11 | 22 | 33 | 44 | 55 | 66 | 77 | 88 | 99 | 110 | 121 | 132 | 143 | 154 |
| 165 | 176 | 187 | 198 | 209 | 220 | 231 | 242 | 253 | 264 | | | | |

Now check whether the total will divide by 11.

How can I use the calculator to check if this number divides by 11?

Note that if the total created by the first nine digits of the ISBN number means that to be divisible by 11, 10 must be added to it, X is used to signify this, not 10, so that only a single digit is used, for example: ISBN 1 874099 26 X.

If the total is divisible by 11, the check digit needs to be zero. Otherwise, this digit should be the smallest number that can be added to 262 so it is divisible by 11.

What is the smallest number that can be added to 262 to make it divisible by 11?

Add 2. 262 + 2 = 264, which is divisible by 11. Check this by looking at the last digit of the ISBN number.

Explain that if the total is not divisible by 11, the computer will reject the order.

Times Table Tactics

pairs

Display the instructions on how to find check digits (Resource sheet 27).

> **How to work out an ISBN check digit**
>
> Write down the first nine digits. **Don't** write down the check digit.
>
> Multiply the first digit by 10, the second digit by 9, the third digit by 8 and so on.
>
> Add these products together.
>
> Now check whether the total will divide by 11.
>
> - If the total is divisible by 11, the check digit needs to be zero.
>
> - If the total is not divisible by 11, the check digit should be the smallest number that can be added to the total to make it divisible by 11. If you need to add 10 to the total, use X.

Children need a book each. They write down the first nine digits of its ISBN number, omitting the last digit. They give this nine-digit number to their partner, who has to work out what the check digit is.

plenary

A complete ISBN number, with its check digit, contains ten single digits. Imagine ISBN numbers where the digits are in ascending or descending order. Would the computer reject or accept orders for those books?

Check these ISBNs together. Ask two or three children to keep a running total on their calculators, as individual children do the relevant multiplications, and to divide the final totals by 11.

> **Something for the children to work on at home**
>
> Find out whether if all ten digits of the ISBN number, including the check digit, are the same, this would produce an acceptable ISBN number.
>
> The answer is yes. For example, the ISBN number 3 333333 33 3 would give the following equation:
>
> $(3 \times 10) + (3 \times 9) + (3 \times 8) + (3 \times 7) + (3 \times 6) + (3 \times 5) + (3 \times 4) + (3 \times 3) + (3 \times 2) + (3 \times 1)$
> $= 3 \times (10 + 9 + 8 + 7 + 6 + 5 + 4 + 3 + 2 + 1)$
> $= 3 \times 55$
>
> If the ISBN consisted of a different digit, the pattern would still be a multiple of 55 (8×55 or 7×55...). As 55 is a multiple of 11, all these results would also be multiples of 11.

0 123456 78 9

or 9 876543 21 0

$(0 \times 10) + (1 \times 9) + (2 \times 8)$
$+ (3 \times 7) + (4 \times 6) +$
$(5 \times 5) + (6 \times 4) + (7 \times 3)$
$+ (8 \times 2) + (9 \times 1) = 165$

165 is divisible by 11

$(9 \times 10) + (8 \times 9)$
$+ (7 \times 8) + (6 \times 7)$
$+ (5 \times 6) + (4 \times 5)$
$+ (3 \times 4) + (2 \times 3)$
$+ (1 \times 2) + (0 \times 1) = 330$

330 is divisible by 11.

Both would be accepted!

Extension: help children set up an ISBN checking system on a computer spreadsheet.

Times Table Tactics

Resource sheets

Resource sheet 1

Name...

×	1	2	3	4	5	6	7	8	9	10
1	1	2	3	4	5	6	7	8	9	10
2	2	4	6	8	10	12	14	16	18	20
3	3	6	9	12	15	18	21	24	27	30
4	4	8	12	16	20	24	28	32	36	40
5	5	10	15	20	25	30	35	40	45	50
6	6	12	18	24	30	36	42	48	54	60
7	7	14	21	28	35	42	49	56	63	70
8	8	16	24	32	40	48	56	64	72	80
9	9	18	27	36	45	54	63	72	81	90
10	10	20	30	40	50	60	70	80	90	100

Name..

×	1	2	3	4	5	6
1	1	2	3	4	5	6
2	2	4	6	8	10	12
3	3	6	9	12	15	18
4	4	8	12	16	20	24
5	5	10	15	20	25	30
6	6	12	18	24	30	36

Name..

1	2	3	4	5	6	7	8	9	10
11	12	13	14	15	16	17	18	19	20
21	22	23	24	25	26	27	28	29	30
31	32	33	34	35	36	37	38	39	40
41	42	43	44	45	46	47	48	49	50
51	52	53	54	55	56	57	58	59	60
61	62	63	64	65	66	67	68	69	70
71	72	73	74	75	76	77	78	79	80
81	82	83	84	85	86	87	88	89	90
91	92	93	94	95	96	97	98	99	100

Name………………………………………………………………

0	1	2	3	4	5	6	7	8	9
10	11	12	13	14	15	16	17	18	19
20	21	22	23	24	25	26	27	28	29
30	31	32	33	34	35	36	37	38	39
40	41	42	43	44	45	46	47	48	49
50	51	52	53	54	55	56	57	58	59
60	61	62	63	64	65	66	67	68	69
70	71	72	73	74	75	76	77	78	79
80	81	82	83	84	85	86	87	88	89
90	91	92	93	94	95	96	97	98	99

Resource sheet 5

How to make a paper clip spinner

Cut out the right spinner for your activity. Use a pencil to hold one end of a paper clip or safety pin in the centre of the spinner. Flick the paper clip or safety pin and see which number it lands on.

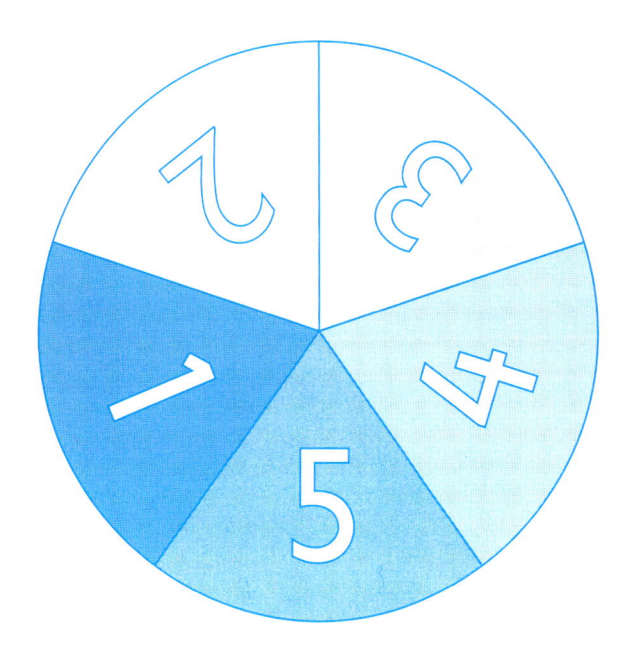

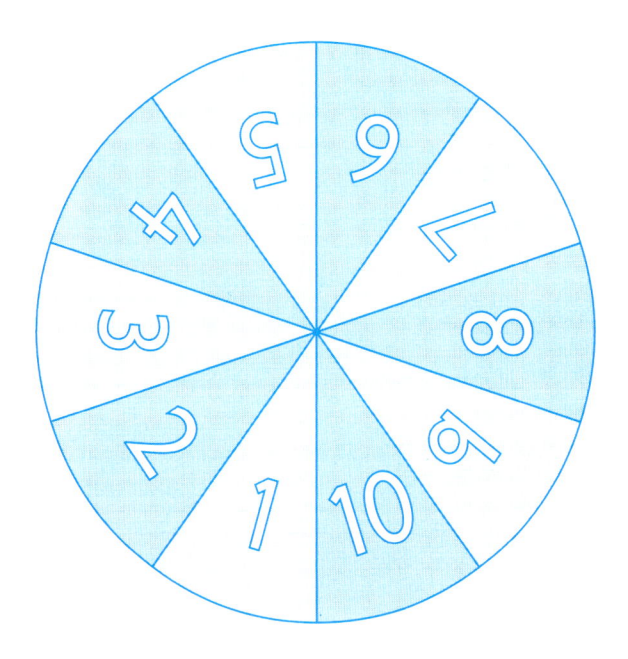

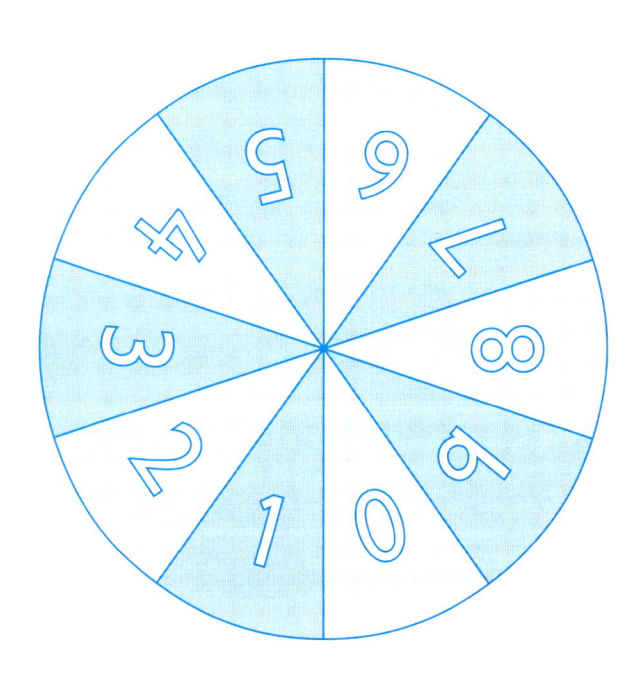

Used with: activities that require either dice or spinners

Times Table Tactics © BEAM Education 2003

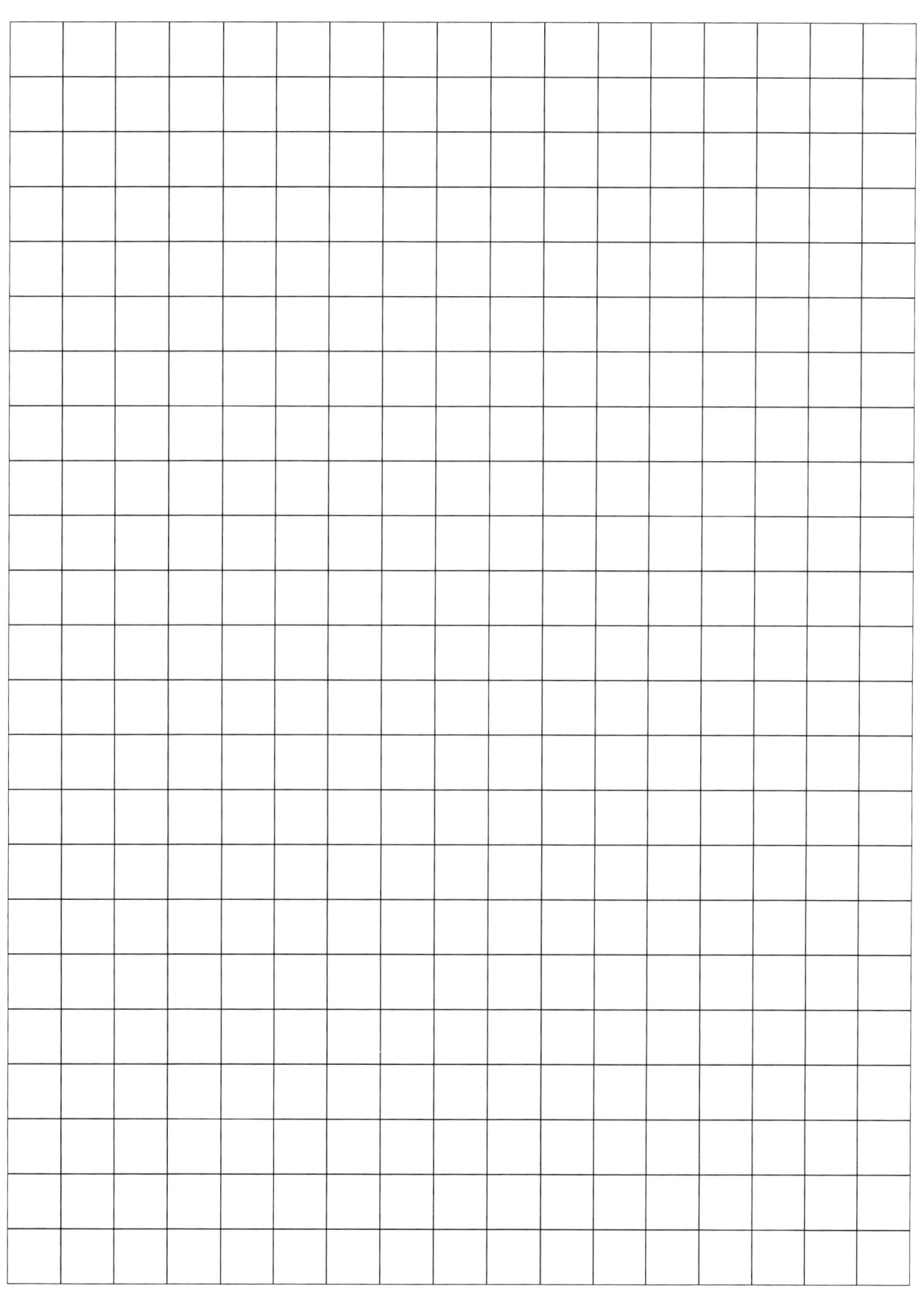

Resource sheet 6

Name..

Resource sheet 7

2	3	4	5
4	6	8	10
6	9	12	15
8	12	16	20
10	15	20	25
12	18	24	30
14	21	28	35
16	24	32	40
18	27	36	45
20	30	40	50

Used with: **Long activities** – Figure out the fours (p26–7), Spot the difference (p36–7)

Times Table Tactics © *BEAM Education 2003*

Resource sheet 8

6	7	8	9
12	14	16	18
18	21	24	27
24	28	32	36
30	35	40	45
36	42	48	54
42	49	56	63
48	56	64	72
56	63	72	81
60	70	80	90

Resource sheet 9

Used with: **Short activities** – Wacky windows (p14)

Times Table Tactics © *BEAM Education 2003*

Resource sheet 10

Used with: Short activities – Wacky windows (p14)

Times Table Tactics © BEAM Education 2003

Resource sheet 11

Name..

number							its multiples							
2	2	4	6	8	10	12	14	16	18	20	22	24	26	28	30
3		3		6		9		12		15		18		21		24		27		30
4			4		8		12		16		20		24		28	
5				5			10			15			20			25			30
6				6			12			18			24			30
7					7				14				21				28
8					8				16				24			
9						9				18					27	
10						10					20					30

Used with Short activities Charting the waves (p15)

Times Table Tactics © BEAM Education 2003

Resource sheet 12

Name..

number						its multiples										
2	30	32	34	36	38	40	42	44	46	48	50	52	54	56	58	60
3																
4																
5																
6																
7																
8																
9																
10																

numbers that don't fit on the grid

Resource sheet 13

Name..

1) Iman works 7 days and gets £9 a day. Emma works 9 days and gets £7 a day. Who earned the most? Why?

2) Ryan earns £7 a week. He gets £9 a week pay rise. Mayur earns £9 a week. He gets £7 a week pay rise. Who earns the most now? Why?

3) How does this diagram show that 9×7 and 7×9 give the same answer?

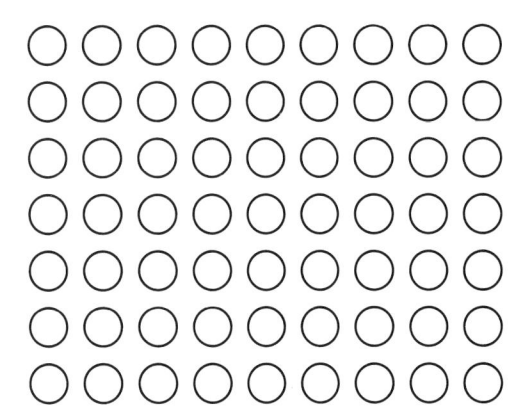

4) How does this diagram show that $9 + 7$ and $7 + 9$ give the same result?

7	9

9	7

5) Is this true or false? $9 \div 7 = 7 \div 9$

6) Is this true or false? $9 - 7 = 7 - 9$

Answers

1) They earned the same: $7 \times 9 = 9 \times 7$
2) Both are now paid the same amount: $7 + 9 = 9 + 7$
3) You can look at the same number of counters in two ways: as seven rows of 9 or nine columns of 7.
4) By symmetry. Rotate either one through 180 degrees, a half turn, and they match. They are also the same length.
5 and 6) False. Multiplication and addition are commutative, but subtraction and division are not.

Used with: Short activities – Money matters (p15)

Times Table Tactics © BEAM Education 2003

Resource sheet 14

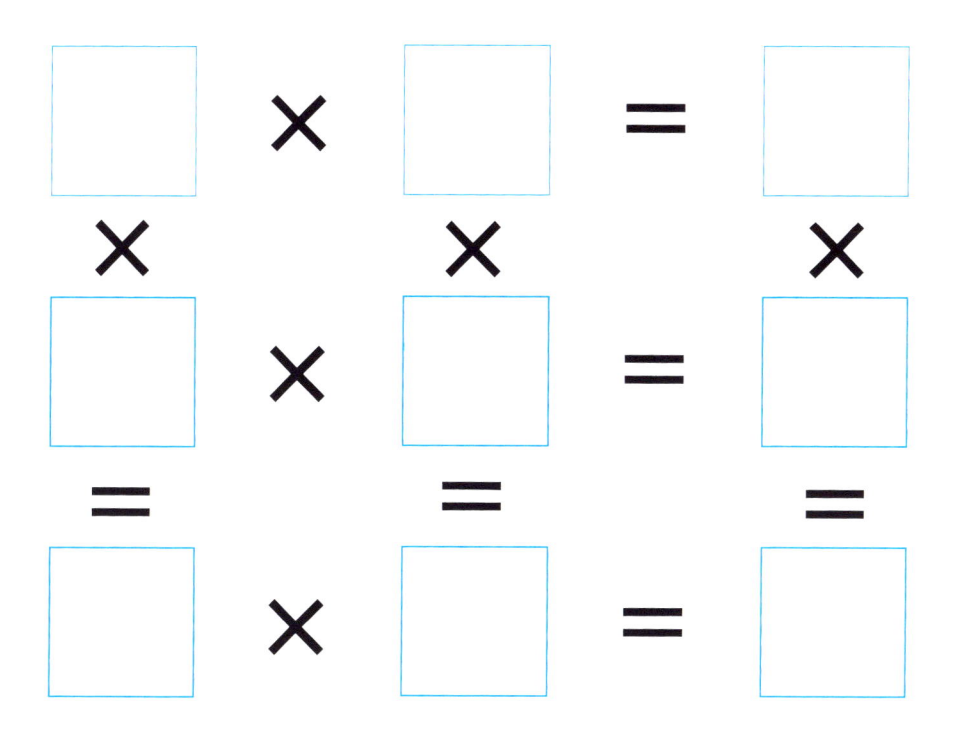

Name..

☐ × ☐ = ☐
×   ×   ×
☐ × ☐ = ☐
=   =   =
☐ × ☐ = ☐

☐ × ☐ = ☐
×   ×   ×
☐ × ☐ = ☐
=   =   =
☐ × ☐ = ☐

Resource sheet 15

Name..

9	7	6	3
3	9	2	7
2	6	1	2
7	3	2	1

8	8	6	4
3	7	2	1
2	5	1	0
4	6	2	4

6	3	1	8
9	4	3	6
3	8	2	4
4	5	2	0

Resource sheet 16

Name..

Grid 1 (5 columns):

1	2	3	4	5
6	7	8	9	10
11	12	13	14	15
16	17	18	19	20
21	22	23	24	25
26	27	28	29	30
31	32	33	34	35
36	37	38	39	40
41	42	43	44	45
46	47	48	49	50
51	52	53	54	55
56	57	58	59	60
61	62	63	64	65
66	67	68	69	70
71	72	73	74	75
76	77	78	79	80
81	82	83	84	85
86	87	88	89	90
91	92	93	94	95
96	97	98	99	100

Grid 2 (6 columns):

1	2	3	4	5	6
7	8	9	10	11	12
13	14	15	16	17	18
19	20	21	22	23	24
25	26	27	28	29	30
31	32	33	34	35	36
37	38	39	40	41	42
43	44	45	46	47	48
49	50	51	52	53	54
55	56	57	58	59	60
61	62	63	64	65	66
67	68	69	70	71	72
73	74	75	76	77	78
79	80	81	82	83	84
85	86	87	88	89	90
91	92	93	94	95	96
97	98	99	100	101	102

Grid 3 (7 columns):

1	2	3	4	5	6	7
8	9	10	11	12	13	14
15	16	17	18	19	20	21
22	23	24	25	26	27	28
29	30	31	32	33	34	35
36	37	38	39	40	41	42
43	44	45	46	47	48	49
50	51	52	53	54	55	56
57	58	59	60	61	62	63
64	65	66	67	68	69	70
71	72	73	74	75	76	77
78	79	80	81	82	83	84
85	86	87	88	89	90	91
92	93	94	95	96	97	98
99	100	101	102	103	104	105

Used with: **Long activities** – Checks and stripes (p24–5)

Times Table Tactics © BEAM Education 2003

Name..

Resource sheet 17

1	2	3	4	5	6	7	8	9
10	11	12	13	14	15	16	17	18
19	20	21	22	23	24	25	26	27
28	29	30	31	32	33	34	35	36
37	38	39	40	41	42	43	44	45
46	47	48	49	50	51	52	53	54
55	56	57	58	59	60	61	62	63
64	65	66	67	68	69	70	71	72
73	74	75	76	77	78	79	80	81
82	83	84	85	86	87	88	89	90
91	92	93	94	95	96	97	98	99

1	2	3	4	5	6	7	8
9	10	11	12	13	14	15	16
17	18	19	20	21	22	23	24
25	26	27	28	29	30	31	32
33	34	35	36	37	38	39	40
41	42	43	44	45	46	47	48
49	50	51	52	53	54	55	56
57	58	59	60	61	62	63	64
65	66	67	68	69	70	71	72
73	74	75	76	77	78	79	80
81	82	83	84	85	86	87	88
89	90	91	92	93	94	95	96
97	98	99	100	101	102	103	104

Used with: **Long activities** – Checks and stripes (p24–5)

Resource sheet 18

Name ....

× 1 =   × 11 =

× 2 =   × 12 =

× 3 =   × 13 =

× 4 =   × 14 =

× 5 =   × 15 =

× 6 =   × 16 =

× 7 =   × 17 =

× 8 =   × 18 =

× 9 =   × 19 =

× 10 =   × 20 =

Resource sheet 19

Name..

	1	2	3	4	5	6	7	8	9	10
× 1										
× 2										
× 10										
× 5										

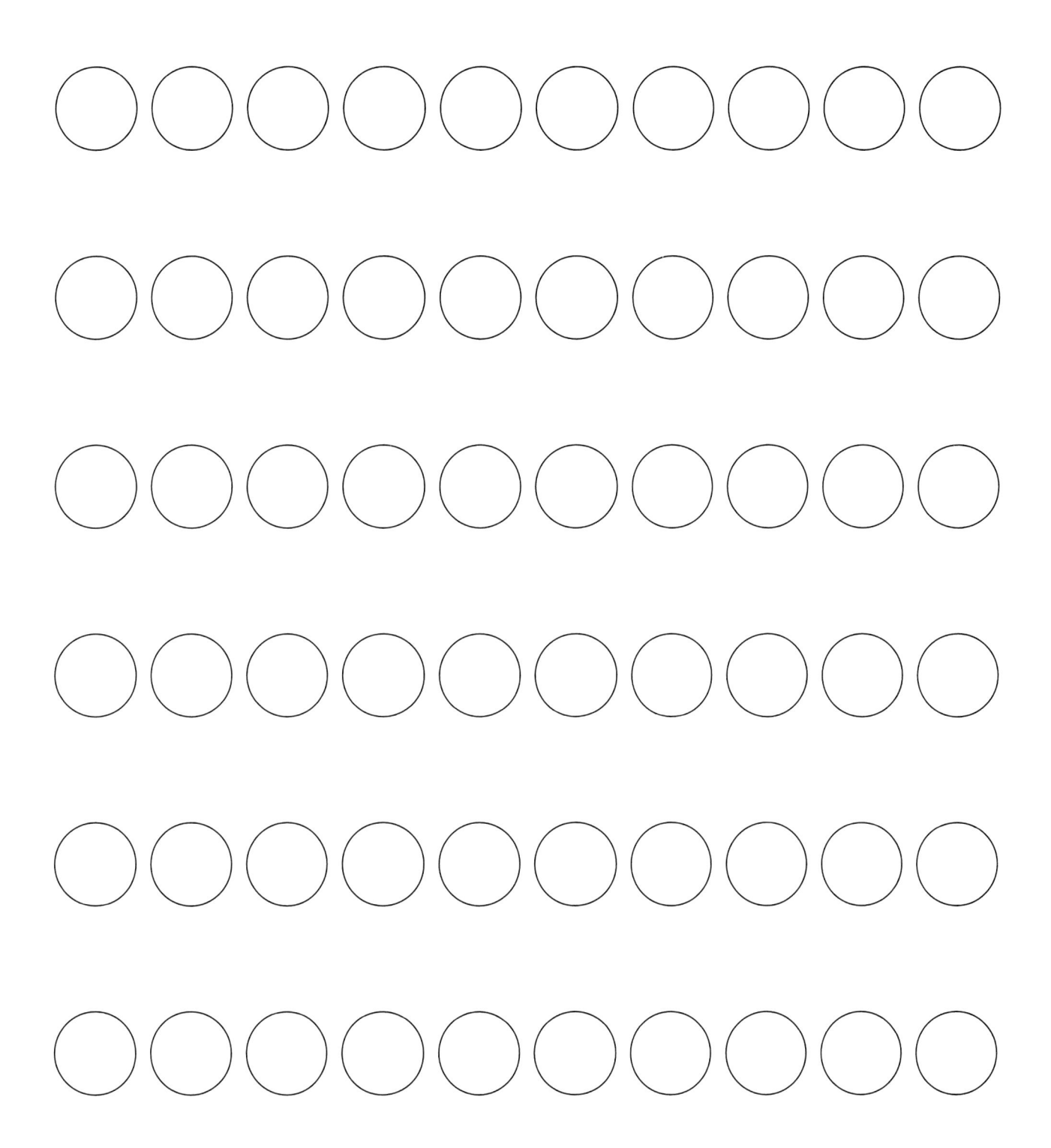

Resource sheet 20

Resource sheet 21

Name..

A $8 \times 9 = (8 \times 5) + (8 \times 4)$
$ = 40 + 32$
$ = 72$

B $8 \times 9 = (8 \times 10) - 8$
$ = 80 - 8$
$ = 72$

C 8×9 is (8, 16, 24, 32, 40, 48, 56, 64, 72)
$8 \times 9 = 72$

D $8 \times 9 = 9 \times 8$
$ = 72$

1 × 1 = 1	2 × 1 = 2	3 × 1 = 3
1 × 2 = 2	2 × 2 = 4	3 × 2 = 6
1 × 3 = 3	2 × 3 = 6	3 × 3 = 9
1 × 4 = 4	2 × 4 = 8	3 × 4 = 12
1 × 5 = 5	2 × 5 = 10	3 × 5 = 15
1 × 6 = 6	2 × 6 = 12	3 × 6 = 18
1 × 7 = 7	2 × 7 = 14	3 × 7 = 21
1 × 8 = 8	2 × 8 = 16	3 × 8 = 24
1 × 9 = 9	2 × 9 = 18	3 × 9 = 27
1 × 10 = 10	2 × 10 = 20	3 × 10 = 30

4 × 1 = 4	5 × 1 = 5
4 × 2 = 8	5 × 2 = 10
4 × 3 = 12	5 × 3 = 15
4 × 4 = 16	5 × 4 = 20
4 × 5 = 20	5 × 5 = 25
4 × 6 = 24	5 × 6 = 30
4 × 7 = 28	5 × 7 = 35
4 × 8 = 32	5 × 8 = 40
4 × 9 = 36	5 × 9 = 45
4 × 10 = 40	5 × 10 = 50

E times table

E × J = E
E × B = JC
E × F = JE
E × G = BC
E × E = BE
E × H = FC
E × A = FE
E × D = GC
E × K = GE

F times table

F × B = F
F × H = BJ
F × A = HG
F × K = AC
F × E = KE
F × C = EK
F × G = CA
F × J = GH
F × F = JB

Resource sheet 24

Name..

Scrambled times tables

A times table	B times table	C times table	D times table
A × K = JC	B × A = A	C × K = GK	D × C = D
A × J = EA	B × F = F	C × F = BJ	D × J = GE
A × G = JA	B × G = G	C × B = D	D × H = GB
A × E = A	B × C = C	C × J = BG	D × F = CG
A × F = HC	B × B = B	C × C = F	D × E = FG
A × B = EC	B × K = K	C × G = C	D × G = E
A × D = HA	B × D = D	C × D = GA	D × D = CA
A × H = BC	B × J = J	C × H = GB	D × A = GD
A × A = BA	B × H = H	C × A = BH	D × K = FA

the code is
A B C D E

F G H J K

the code is
A B C D E

F G H J K

the code is
A B C D E

F G H J K

Resource sheet 25

Name ....

15	15	×	5	×	3	×	×	×	×	×	×	×	×	×	1
14	14	7	×	×	×	×	2	×	×	×	×	×	×	1	×
13	13	×	×	×	×	×	×	×	×	×	×	×	1	×	×
12	12	6	4	3	×	2	×	×	×	×	1	×	×	×	×
11	11	×	×	×	×	×	×	×	×	×	1	×	×	×	×
10	10	5	×	×	2	×	×	×	×	1	×	×	×	×	×
9	9	×	3	×	×	×	×	1	×	×	×	×	×	×	×
8	8	4	×	2	×	×	1	×	×	×	×	×	×	×	×
7	7	×	×	×	×	1	×	×	×	×	×	×	×	×	×
6	6	3	2	×	×	1	×	×	×	×	×	×	×	×	×
5	5	×	×	×	1	×	×	×	×	×	×	×	×	×	×
4	4	2	×	1	×	×	×	×	×	×	×	×	×	×	×
3	3	×	1	×	×	×	×	×	×	×	×	×	×	×	×
2	2	1	×	×	×	×	×	×	×	×	×	×	×	×	×
1	1	×	×	×	×	×	×	×	×	×	×	×	×	×	×
÷	1	2	3	4	5	6	7	8	9	10	11	12	13	14	15

Used with: **Long activities** — All or nothing (p54–5)

Times Table Tactics © BEAM Education 2003

Resource sheet 26

Question cards

8 1	9 1
8 2	9 2
8 3	9 3
8 4	9 4
8 5	9 5
8 6	9 6
8 7	9 7
8 8	9 8
8 9	9 9

Answer cards

8	9
16	18
24	27
32	36
40	45
48	54
56	63
64	72
72	81

Photocopy the question and answer cards onto different coloured paper, to help children distinguish them.

How to work out an ISBN check digit

Write down the first nine digits. **Don't** write down the check digit.

Multiply the first digit by 10, the second digit by 9, the third digit by 8 and so on.

Add these products together.

Now check whether the total will divide by 11.

- If the total is divisible by 11, the check digit needs to be zero.

- If the total is not divisible by 11, the check digit should be the smallest number that can be added to the total to make it divisible by 11. If you need to add 10 to the total, use X.